Recognized as an
American National Standard (ANSI)

IEEE Std 1015-1997

IEEE Recommended Practice for Applying Low-Voltage Circuit Breakers Used in Industrial and Commercial Power Systems

Sponsor

Power Systems Protection Committee
of the
Industrial and Commercial Power Systems Department
of the
IEEE Industry Applications Society

Approved 10 February 1997

IEEE Standards Board

Approved 18 June 1997

American National Standards Institute

Abstract: Information is provided for selecting the proper circuit breaker for a particular application. This recommended practice helps the application engineer specify the type of circuit breaker, ratings, trip functions, accessories, acceptance tests, and maintenance requirements. It also discusses circuit breakers for special applications, e.g., instantaneous only and switches. In addition, it provides information for applying circuit breakers at different locations in the power system, and for protecting specific components. Guidelines are also given for coordinating combinations of line-side and load-side devices.
Keywords: circuit breaker, circuit breaker evaluation, insulated case, insulated-case circuit breaker, low-voltage circuit breaker, low-voltage power circuit breaker, low-voltage protection, low-voltage protective device, molded case, molded-case circuit breaker, overcurrent protection, power circuit breaker, rating, testing

Grateful acknowledgment is made to the following organizations for having granted permission to reprint material in this document as listed below:

American National Standards Institute, 11 West 42nd Street, New York, NY 10036, for source data from ANSI C37.16-1988 used in Tables 3-8, 3-9, 3-10, 3-12, 3-14, 4-5, and 4-6, and for source data from ANSI C37.50-1989 used in Table 3-20.

National Fire Protection Association (NFPA), 1 Batterymarch Park, P.O. Box 9101, Quincy, MA 02269-9101, for source data from NEC Section 240-6 used in Table 3-11.

Siemens Energy & Automation, Inc., Switchgear & Motor Control, P.O. Box 29503, Raleigh, NC 27626-0503, for Figures 1-2 and 6-1 (right photo).

Square D Company, Distribution Equipment Business, Box 3069, 3700 Sixth Street S.W., Cedar Rapids, IA 52406, for Figures 1-1, 6-1 (left photo), and 6-2.

Underwriters Laboratory, 333 Pfingsten Road, Northbrook, IL 60062-2096, for source data from UL 489-1991 used in Tables 3-2, 3-4, 3-6, 3-13, 3-17, 3-18, and 3-19; and Figure 3-1, and for source data from the UL Recognized Component Directory, vol. 1, 1996 used in Table 4-1.

U.S. Government Printing Office, Washington, DC 20037, for information from the Code of Federal Regulations, Title 30, Chapter 1, Section 75.601-1, used in Table 6-1.

Westinghouse (*now* Cutler-Hammer), for source data used in Tables 3-1, 3-3, 3-15, 3-16, 3-21, and 3-22.

First Printing
October 1997
SH94473

The Institute of Electrical and Electronics Engineers, Inc.
345 East 47th Street, New York, NY 10017-2394, USA

ISBN 1-55937-867-0

Introduction

(This introduction is not a part of IEEE Std 1015-1997, IEEE Recommended Practice for Applying Low-Voltage Circuit Breakers Used in Industrial and Commercial Power Systems.)

This recommended practice is a result of a 13-year effort by members of the Low-Voltage Protection Subcommittee of the Power System Protection Committee of The IEEE Industrial Applications Society. This publication, known as the IEEE Blue Book, will be added to the present IEEE Color Book series sponsored by the IAS Industrial and Commercial Power Systems Department.

In the past it was often very difficult and time consuming for an engineer to decide what type of low-voltage circuit breaker to use for a particular application, and how to apply the circuit breaker. The application guidelines and tables were dispersed over many different documents and standards. There was also confusion caused by biases based on the competitiveness between the different product classes. Thus, a working group was formed among the members of the Low-Voltage Protection Subcommittee to write a recommended practice that provides a comprehensive reference source for the selection and application of low-voltage circuit breakers. This recommended practice includes a comparison between the standards of low-voltage power circuit breakers and molded-case circuit breakers so that an engineer can make better, more informed choices. Pertinent tables have been extracted from other standards to provide the basis for the selection and application guidelines. In addition, specific application examples are provided.

To the many members of the working group who wrote and developed this recommended practice, we owe a debt of gratitude for the many hours that were spent writing and editing the IEEE Blue Book chapters. These individuals deserve our many thanks for their excellent contributions.

This IEEE Recommended Practice serves as a companion publication to the following other Recommended Practices prepared by the IEEE Industrial and Commercial Power Systems Department:

— IEEE Std 141-1993, IEEE Recommended Practice for Electric Power Distribution for Industrial Plants (IEEE Red Book).

— IEEE Std 142-1991, IEEE Recommended Practice for Grounding of Industrial and Commercial Power Systems (IEEE Green Book).

— IEEE Std 241-1990, IEEE Recommended Practice for Electric Power Systems in Commercial Buildings (IEEE Gray Book).

— IEEE Std 242-1986, IEEE Recommended Practice for Protection and Coordination of Industrial and Commercial Power Systems (IEEE Buff Book).

— IEEE Std 399-1990, IEEE Recommended Practice for Industrial and Commercial Power Systems Analysis (IEEE Brown Book).

— IEEE Std 446-1995, IEEE Recommended Practice for Emergency and Standby Power Systems for Industrial and Commercial Applications (IEEE Orange Book).

— IEEE Std 493-1990, IEEE Recommended Practice for the Design of Reliable Industrial and Commercial Power Systems (IEEE Gold Book).

— IEEE Std 602-1996, IEEE Recommended Practice for Electric Systems in Health Care Facilities (IEEE White Book).

— IEEE Std 739-1995, IEEE Recommended Practice for Energy Management in Industrial and Commercial Facilities (IEEE Bronze Book).

— IEEE Std 1100-1992, IEEE Recommended Practice for Powering and Grounding Sensitive Electronic Equipment (IEEE Emerald Book).

The Blue Book Working Group for the 1997 edition had the following membership:

Keith R. Cooper, *Chair* **William M. Hall,** *Vice Chair*
George D. Gregory, *Secretary*

Chapter 1: Overview—**Keith R. Cooper,** Chair

Chapter 2: Definitions and acronyms—**George D. Gregory,** Chair

Chapter 3: Rating and testing—**Raymond D. Valentine,** *Chair;*
Joseph J. Andrews, Bruce G. Bailey, William M. Hall, Ed Larsen, Bill May, Bill Reardon, Dean Sigmon

Chapter 4: Specific applications—**James W. Brosnahan,** *Co-Chair;* **Shaun P. Slattery,** *Co-Chair;*
Joseph J. Andrews, Bruce G. Bailey, David S. Baker, Rene D. Castenschiold, Ray M. Clark, Keith R. Cooper, William M. Hall, Walter Huening, Donald H. McCullough, Raymond D. Valentine

Chapter 5: Selective coordination of low-voltage circuit breakers with other protective devices—**Jay Fischer,** *Chair;*
Jack Alacchi, Vincent Saporita, Ralph Young

Chapter 6: Special-purpose circuit breakers—**George D. Gregory,** *Chair;*
Ed Larsen

Chapter 7: Acceptance and maintenance requirements—**Daniel J. Love,** *Co-Chair;* **Donald H. McCullough,** *Co-Chair;* **Alan C. Pierce,** *Co-Chair;*
Joseph J. Andrews, Kenneth W. Carrick, Bill Reardon, Robert Simpson

Others who contributed to the development of this document are as follows:

L. Guy Jackson Russell Ohlson Raymond O. D. Whitt

The following persons were on the balloting committee:

Joseph J. Andrews
Bruce G. Bailey
David S. Baker
James W. Brosnahan
Kenneth W. Carrick
Rene D. Castenschiold
Ray M. Clark
Keith R. Cooper
Jay Fischer
Edgar O. Galyon
George D. Gregory
William M. Hall
L. Guy Jackson
James R. Jones
Shankar Lakhavani
Daniel J. Love
Donald H. McCullough
Alan C. Pierce
Vincent Saporita
Stephen Schaffer
Shaun P. Slattery
Henry Smith
Raymond D. Valentine
Ralph H. Young

When the IEEE Standards Board approved this recommended practice on 10 February 1997, it had the following membership:

Donald C. Loughry, *Chair* **Richard J. Holleman,** *Vice Chair*
Andrew G. Salem, *Secretary*

Gilles A. Baril
Clyde R. Camp
Joseph A. Cannatelli
Stephen L. Diamond
Harold E. Epstein
Donald C. Fleckenstein
Jay Forster*
Donald N. Heirman
Ben C. Johnson
E. G. "Al" Kiener
Joseph L. Koepfinger*
Stephen R. Lambert
Lawrence V. McCall
L. Bruce McClung
Marco W. Migliaro
Mary Lou Padgett
John W. Pope
Jose R. Ramos
Arthur K. Reilly
Ronald H. Reimer
Gary S. Robinson
Ingo Rüsch
John S. Ryan
Chee Kiow Tan
Leonard L. Tripp
Howard L. Wolfman

*Member Emeritus

Also included are the following nonvoting IEEE Standards Board liaisons:

Satish K. Aggarwal
Alan H. Cookson
Chester C. Taylor

Valerie E. Zelenty
IEEE Standards Project Editor

Contents

IEEE

IEEE Recommended Practice for Applying Low-Voltage Circuit Breakers Used in Industrial and Commercial Power Systems

Chapter 1 Overview

1.1 Scope

This recommended practice provides information for selecting the proper circuit breaker for a particular application. This recommended practice helps the application engineer specify the type of circuit breaker, ratings, trip functions, accessories, acceptance tests, and maintenance requirements. It also discusses circuit breakers for special applications, e.g., instantaneous only and switches. In addition, it provides information for applying circuit breakers at different locations in the power system, and for protecting specific components. Guidelines are given for coordinating combinations of line-side and load-side devices. Acceptance testing and maintenance guidelines are provided so that reliable operation can be verified and maintained.

This recommended practice does not cover the selection and application of circuit breakers such as ground-fault circuit interrupters (GFCIs), marine circuit breakers, definite purpose circuit breakers, and supplementary circuit protectors. This recommended practice also does not discuss dc applications.

1.2 Two classifications of breakers

There are two main classifications of low-voltage circuit breakers: molded-case circuit breakers and low-voltage power circuit breakers. Within the molded-case circuit breaker classification, there is another type of circuit breaker called the insulated-case circuit breaker. The construction and characteristics of these three types will be discussed. Throughout the balance of this recommended practice these devices will be referred to as the following:

— MCCB molded-case circuit breaker
— ICCB insulated-case circuit breaker
— LVPCB low-voltage power circuit breaker

Each one of these circuit breakers has different design characteristics, and in many cases, different application requirements.

This recommended practice compares the circuit breakers so that the power systems engineer can decide which one is best suited for a particular application. In addition, it discusses ratings, such as overload, short-time, and interrupting capabilities. Protection requirements depend on the circuit breaker location in the power system as well as the type of equipment that is being protected. Examples for different types of equipment and circuit locations are discussed in this recommended practice.

MCCBs are tested and rated in accordance with UL 489-1991.[1] Their current-carrying parts, mechanisms, and trip devices are completely contained within a molded case of insulating material. The cover and base of smaller MCCBs are designed so that the MCCBs cannot be opened for maintenance purposes. The main contacts of MCCBs cannot be removed; however, some MCCBs are available with field-installable accessories. MCCBs are available in stationary or plug-in construction with circuit-breaker enclosures that can be flush or surface mounted. They are available in a large number of continuous-current and interrupting ratings. The smaller continuous-current ratings are equipped with thermal-magnetic or magnetic only trip units. Larger sizes are also available with thermal-magnetic or electronic (static) trip devices.

ICCBs are also tested and rated in accordance with UL 489-1991. As with other MCCBs, ICCB current-carrying parts, mechanisms, and trip units are contained within a molded case of insulating material. The case is designed so it can be opened for inspection of contacts and arc chutes and for limited maintenance. Most manufacturers offer designs that permit replacement of accessories, and some designs permit replacement of the main contacts. ICCBs are available in both stationary and drawout construction. They are generally characterized by a stored energy mechanism, larger frame sizes, and higher short-time withstand ratings than MCCBs. Electronic trip units are standard.

LVPCBs are tested and rated according to the following standards:

- ANSI Std C37.16-1988
- ANSI Std C37.17 -1979
- ANSI Std C37.50-1989
- IEEE Std C37.13-1990
- UL 1558-1993

LVPCBs are generally characterized by physically large frame sizes, drawout construction, and the highest short-time withstand ratings of all the types of low-voltage circuit breakers. When the circuit breaker is removed from its enclosure, the current-carrying parts and operating parts are accessible for inspection, maintenance, and replacement purposes. Electromechanical trip units were used in the circuit breakers prior to the early 1970s. However, electronic trip units are used in new LVPCBs and are available as upgrades for older units.

[1]Information on references can be found in 1.7.

1.3 Description of a molded-case circuit breaker

Figure 1-1 is a cutaway view of a typical MCCB. Letters are used to indicate the various elements of the circuit breaker, with a description listed in the legend. This typical circuit breaker operates using an over-center toggle, quick-make-quick-break mechanism. This mechanism is operated manually to the ON (closed) and OFF (open) positions using the handle. The quick-make-quick-break action ensures that the speed at which the breaker contacts are open or closed is independent of the speed at which the handle is moved. This toggle mechanism is also trip-free, which means that the circuit breaker cannot be prevented from tripping by holding or locking the handle in the ON position. When the circuit breaker trips open automatically, the handle will assume either an intermediate position between ON and OFF or the OFF position. If the handle moves to the intermediate position, it must be manually moved slightly past the OFF position to reset the mechanism. Other instructions for resetting a particular circuit breaker after it trips should be marked on the circuit breaker and/or indicated on the equipment where the circuit breaker is installed.

1.4 Description of a low-voltage power circuit breaker

Figure 1-2 is a view of a partially disassembled, manually operated, drawout LVPCB. The open construction permits access to the circuit-breaker parts for maintenance and parts replacement. Numbers are used to indicate the various elements of the circuit breaker. A description of each element is listed in the legend. The following is a description of the operation of the circuit breaker starting with the open position. The circuit breaker condition "open" is indicated on the face of the circuit breaker. In order to close the circuit breaker, a spring mechanism must be charged. The springs are charged by pulling down and releasing the manual spring charging handle. The spring condition "charged" is indicated on the face of the circuit breaker. The circuit breaker is manually closed by depressing the close (push-to-close) hood. The circuit breaker condition "closed" is indicated on the face of the circuit breaker. The circuit breaker is opened manually by depressing the open (push-to-trip) lever or automatically by the operation of the trip unit.

The drawout circuit breaker has three separate positions: "connected," "test," and "disconnected." A racking crank is used to move the drawout circuit breaker to each position in the circuit breaker compartment. The circuit breaker's contacts are only connected to the external power circuit in the connected position. If the circuit breaker is closed in the test position, there is no effect to the external power circuit. Interlocking prevents moving a closed circuit breaker between these positions or closing it in other than the connected or test position. Further interlocking prevents inserting circuit breakers of the wrong frame size into a compartment. Primary disconnects and optional secondary disconnects automatically complete the power circuit in the connected position and control circuits in the connected and test positions, respectively.

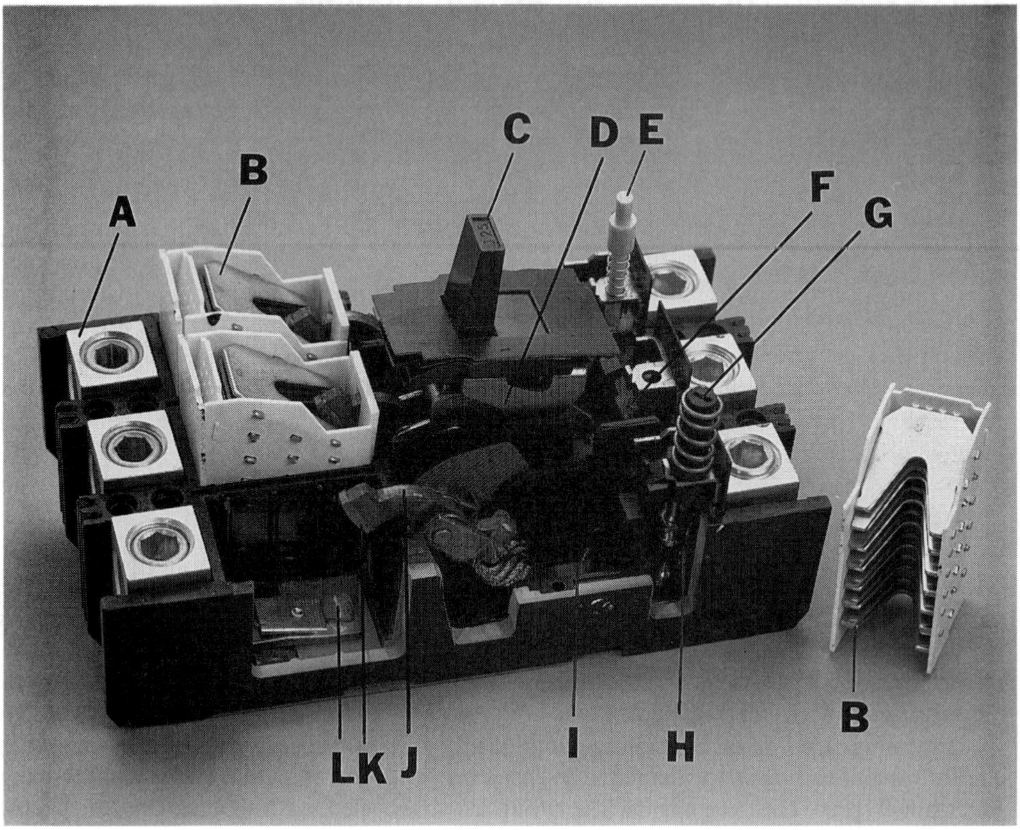

LEGEND

A.	Wire connector	G.	Instantaneous trip level adjustment
B.	De-ionizing arc stack	H.	Electro-magnet
C.	Handle	I.	Bimetal
D.	Operating mechanism	J.	Moving arm
E.	Test trip actuator	K.	Moving contact
F.	Common-trip bar	L.	Stationary contact

Source: Square D Company.

Figure 1-1—Cutaway view of a typical MCCB

LEGEND

1. Manual spring charging handle
2. Open (push-to-trip) lever
3. Close (push-to-close) hood
4. Racking crank access opening & interlock
5. Racking (drawout) mechanism
6. Drawout interlocks
7. Breaker frame size interlock
8. Primary disconnect finger assembly
9. Main contacts
10. Arcing contacts
11. Arc chute
12. Interphase barrier
13. Electronic overcurrent trip device
14. Current sensors (on rear—not shown)
15. Voltage sensor (optional)
16. Breaker display unit (current, voltage, and power measurements) (optional)

Source: Siemens Energy & Automation, Inc.

Figure 1-2—Low-voltage ac power circuit breaker—drawout type (shown partially disassembled to show internal features)

1.5 Document organization

This recommended practice is organized in the following manner:

Chapter 2: Definitions and acronyms.
All of the major terms used in other chapters are defined in this chapter. For a listing of additional electrical definitions, refer to IEEE Std 100-1996 and IEEE Std C37.100-1992.

Chapter 3: Rating and testing.
This chapter summarizes the application parts of recognized low-voltage circuit breaker standards. For more details in a particular application, the engineer is encouraged to refer to the complete standard for amplification and a more complete discussion.

By understanding the differences in standards, an engineer can make a better decision about which type of breaker should be used for a particular application. In addition, the National Electrical Code® (NEC®) (NFPA 70-1996), Section 110-3(b), requires that a circuit breaker be applied according to the information listed in the standards. Thus it is important that the engineer know and understand the standards so that proper application procedures are followed.

Chapter 4: Specific applications.
This chapter provides a systematic procedure for selecting and applying MCCBs, ICCBs, and LVPCBs at various locations in a power system. The applications covered in Chapter 4 include the following:

- Service entrances
- Mains (buses and busway, feeder and branch protective devices, and line-side transformers)
- Bus tie
- Feeders and branch circuits (cable and busway)
- Circuit breakers in series combinations
- Motors (individual and grouped)
- Transformers
- Capacitors and capacitor banks
- Generators
- Switchboards and panelboards
- Motor control centers and starters

Chapter 5: Selective coordination of low-voltage circuit breakers with other protective devices.
Conflicting objectives normally occur between protection and selective coordination. The objective of protection is to minimize the damage by removing the overload or short circuit as quickly as possible. However, the objective of selective coordination is to disconnect a minimum amount of equipment from the power system. Coordination is obtained by selecting the appropriate type of circuit breaker, trip characteristics, and trip settings so that only the circuit breaker closest to the overload or short-circuit condition clears the problem.

In order to obtain selective coordination over the entire range of available short-circuit current, delayed operation of the line-side circuit breaker is necessary to allow the load-side circuit breaker to clear a fault. The additional time delay of the line-side device can increase the extent of the damage to circuit components. However, it is often necessary to make a compromise in protection to obtain coordination, within the limits of NEC and ANSI protection requirements. Coordination between devices is often sacrificed when continuity of service is not critical because of the additional equipment cost. This chapter shows how to coordinate different combinations of devices, and provides typical settings.

Chapter 6: Special-purpose circuit breakers.
This chapter discusses application of the following special-purpose circuit breakers:

- Instantaneous only
- Mine duty
- Current limiting
- Molded-case switch
- Integrally fused

Chapter 7: Acceptance and maintenance requirements.
This chapter is divided into the following two time periods:

- Acceptance testing that should be performed *before* placing a circuit breaker into service
- Maintenance guidelines that should be followed *after* a circuit breaker is placed into service

Acceptance testing and maintenance are required to ensure that the circuit breaker will perform satisfactorily to the full extent of the manufacturer's ratings. In addition, proper maintenance is required for continued, reliable operation.

Acceptance testing of circuit breakers is the initial testing on a breaker before it is placed into service to verify that the circuit breaker will perform properly. A circuit breaker may be damaged during shipment, or defective components may be present. This chapter provides instructions for performing and documenting test results. Some of the acceptance test criteria discussed are insulation resistance, electrical operations at selected settings, mechanical operation, and auxiliary functions. Acceptance test data sheets are provided.

Maintenance procedures are important to power system reliability and safety. Proper maintenance ensures the correct operation of circuit breakers to interrupt an overload or fault condition. The operating performance of a circuit breaker may deteriorate with time due to contamination by pollutants in the atmosphere, excessive use, or lack of switching or tripping. Welding in the main contact may occur during a short circuit if the contacts are in a deteriorated condition due to excessive use. In addition, the performance characteristics of other components in the breaker, such as contact pressure springs, may be degraded for some other reason such as overheating. This chapter discusses proper maintenance procedures and schedules that should be followed to maintain reliable operation.

1.6 Summary

This recommended practice identifies the differences between the two classifications of low-voltage circuit breakers and provides information for making a decision for selecting the best one for a particular situation. In addition, this recommended practice gives information for applying circuit breakers at different locations in the power system, and for protecting specific components. Guidelines are also given for coordinating combinations of line-side and load-side devices. Acceptance testing and maintenance guidelines are given so that reliable operation can be verified and maintained.

1.7 References

This chapter shall be used in conjunction with the following publications:

ANSI C37.16-1988 (Reaff 1995), American National Standard for Switchgear—Low-Voltage Power Circuit Breakers and AC Power Circuit Protectors—Preferred Ratings, Related Requirements, and Application Recommendations.[2]

ANSI C37.17-1979 (Reaff 1988), American National Standard for Trip Devices for AC and General Purpose DC Low-Voltage Power Circuit Breakers.

ANSI C37.50-1989 (Reaff 1995), American National Standard for Switchgear—Test Procedures for Low-Voltage AC Power Circuit Breakers Used in Enclosures.

IEEE Std 100-1996, IEEE Standard Dictionary of Electrical and Electronics Terms.[3]

IEEE Std C37.13-1990 (Reaff 1995), IEEE Standard for Low-Voltage AC Power Circuit Breakers Used in Enclosures (ANSI).

IEEE Std C37.100-1992, IEEE Standard Definitions for Power Switchgear (ANSI).

NFPA 70-1996, National Electrical Code® (NEC®).[4]

UL 489-1991, Molded-Case Circuit Breakers and Circuit-Breaker Enclosures (DoD).[5]

UL 1558-1993, Metal-Enclosed Low-Voltage Power Circuit Breaker Switchgear.

---

[2]ANSI Publications are available from the Sales Department, American National Standards Institute, 11 West 42nd Street, 13th Floor, New York, NY 10036, USA.

[3]IEEE publications are available from the Institute of Electrical and Electronics Engineers, 445 Hoes Lane, P.O. Box 1331, Piscataway, NJ 08855-1331, USA.

[4]The NEC is available from Publications Sales, National Fire Protection Association, 1 Batterymarch Park, P.O. Box 9101, Quincy, MA 02269-9101, USA. It is also available from the Institute of Electrical and Electronics Engineers, 445 Hoes Lane, P.O. Box 1331, Piscataway, NJ 08855-1331, USA.

[5]UL standards are available from Global Engineering, 1990 M Street NW, Suite 400, Washington, DC, 20036, USA.

Chapter 2
Definitions and acronyms

2.1 Definitions

Definitions of terms used in this recommended practice are noted in this chapter. Definitions taken from other references identify the source in parentheses following the definition.

2.1.1 adjustable: (as applied to circuit breakers) A qualifying term indicating that the circuit breaker can be set to trip at various values of current and/or time within a predetermined range. (NFPA 70-1996)[1]

2.1.2 alarm switch: An auxiliary switch that actuates a signaling device upon the automatic opening of the circuit breaker with which it is associated.

2.1.3 auxiliary switch: A switch that is mechanically operated by the main switching device for signaling, interlocking, or other purposes.

NOTE—Auxiliary switch contacts are classified as a, b, aa, bb, LC, etc., for the purpose of specifying definite contact positions with respect to the main device.

2.1.4 available short-circuit current: (at a given point in a circuit) The maximum current that the power system can deliver through a given circuit to any negligible-impedance short circuit applied at the given point, or at any other point that will cause the highest current to flow through the given point. *See also:* **prospective fault current.**

2.1.5 circuit breaker: A device designed to open and close a circuit by nonautomatic means, and to open the circuit automatically on a predetermined overcurrent without damage to itself when properly applied within its rating. (NFPA 70-1996)

2.1.6 continuous current: A current that is expected to continue for three hours or more.

2.1.7 coordination: Properly localizing a fault condition to restrict outages to the equipment affected, accomplished by a choice of selective fault-protective devices. (NFPA 70-1996) *Syn:* **selectivity.**

2.1.8 current-limiting circuit breaker: A circuit breaker that does not use a fusible element and that when operating within its current-limiting range limits the let-through I^2t to a value less than the I^2t of a 1/2 cycle wave of the symmetrical prospective current. (UL 489-1991) *See also:* **current-limiting overcurrent protective device.**

2.1.9 current-limiting overcurrent protective device: A device that, when interrupting currents in its current-limiting range, will reduce the current flowing in the faulted circuit to a magnitude substantially less than that obtainable in the same circuit if the device were

[1]Information on references can be found in 2.3.

replaced with a solid conductor having comparable impedance. (NFPA 70-1996) *See also:* **current-limiting circuit breaker.**

2.1.10 drawout-mounted circuit breaker: An assembly of a circuit breaker together with a supporting structure so constructed that the breaker is supported and can be moved between the main circuit connected and disconnected positions without unbolting connections or mounting supports. The stationary portion of the drawout assembly includes self-supporting primary circuit terminals and may include an interlocking means that permits movement of the breaker between the main circuit connected and disconnected positions only when the breaker contacts are in the open position. (NEMA AB 1-1993)

2.1.11 dynamic impedance: The arc impedance introduced into a circuit by the opening of the circuit-breaker contacts during current interruption. (NEMA AB 1-1993)

2.1.12 electrical operator: A controlling device that is used to open, close, and reset a circuit breaker. Editor's note: The term motor operator is sometimes used when the operating device is a motor. (UL 489-1991)

2.1.13 electronic trip unit: A self-contained portion of a circuit breaker that senses the condition of the circuit breaker electronically and that actuates the mechanism that opens the circuit breaker contacts automatically.

NOTE—Where the term *electronic trip unit* is used in this book, the alternate terms *solid state trip unit* and *static trip unit* are commonly used in literature.

2.1.14 frame: As applied to circuit breakers, that portion of an interchangeable trip unit circuit breaker remaining when the interchangeable trip unit is removed. (UL 489-1991)

2.1.15 frame size: A term applied to a group of circuit breakers of similar physical configuration. Frame size is expressed in amperes and corresponds to the largest ampere rating available in the group. The same frame size designation may be applied to more than one group of circuit breakers. (UL 489-1991)

2.1.16 ground-fault delay: An intentional time delay in the tripping of a circuit breaker when a ground fault occurs. (NEMA AB 1-1993)

2.1.17 ground-fault pickup: The nominal value of the ground fault current at which the ground fault delay function is initiated. (NEMA AB 1-1993)

2.1.18 ground-fault protection of equipment: A system intended to provide protection of equipment from damaging line-to-ground fault currents by operating to cause a disconnecting means to open all ungrounded conductors of the faulted circuit. This protection is provided at current levels less than those required to protect conductors from damage through the operation of a supply circuit overcurrent device. (NFPA 70-1996)

2.1.19 I^2t: An expression related to the energy available as a result of current flow, meaningful only for adiabatic conditions. With respect to circuit breakers, the expression refers to the I^2t between the initiation of fault current and the clearing of the circuit. The defining equation is: $I^2t = I^2(t)$ dt over the stated period, in units of amperes-squared seconds. (UL 489-1991) (NEMA AB 1-1993)

2.1.20 instantaneous pickup: The nominal value of current at which an adjustable circuit breaker is set to trip instantaneously. (NEMA AB 1-1993)

2.1.21 instantaneous trip: (as applied to circuit breakers) A qualifying term indicating that no delay is purposely introduced in the tripping action of the circuit breaker. (NFPA 70-1996)

2.1.22 instantaneous-trip-only circuit breaker: A circuit breaker intended to provide short-circuit protection only. Although acting instantaneously under short-circuit conditions, instantaneous trip breakers shall be permitted to include a transient dampening action to ride through initial motor transients. (NEMA AB 1-1993)

2.1.23 insulated-case circuit breaker (ICCB): A circuit breaker that is assembled as an integral unit in a supporting and enclosing housing of insulating material and with a stored energy mechanism.

2.1.24 integrally-fused circuit breaker: A circuit breaker in which coordinated fuses are connected in series with the release (trip) elements of the circuit breaker and are mounted within the housing of the circuit breaker. (NEMA AB 1-1993)

2.1.25 interrupting rating: The highest current at rated voltage that a device is intended to interrupt under standard test conditions. (NFPA 70-1996)

2.1.26 inverse time: (as applied to circuit breakers) A qualifying term indicating that there is purposely introduced a delay in the tripping action of the circuit breaker, which delay decreases as the magnitude of the current increases. (NFPA 70-1996)

2.1.27 Ip: *See:* **peak current.**

2.1.28 long-time delay: An intentional time delay in the overload tripping of an adjustable circuit breaker's inverse time characteristic. (NEMA AB 1-1993)

2.1.29 long-time pickup: The current at which the long-time delay function is initiated. (NEMA AB 1-1993)

2.1.30 low-voltage power circuit breaker (LVPCB): A mechanical switching device, capable of making, carrying, and breaking currents under normal circuit conditions and also, making and carrying for a specified time and breaking currents under specified abnormal circuit conditions such as those of short-circuit. Rated 1000 V ac or below, or 3000 V dc and below, but not including molded-case circuit breakers. (IEEE Std C37.100-1992)

2.1.31 molded-case circuit breaker (MCCB): A circuit breaker that is assembled as an integral unit in a supporting and enclosing housing of insulating material. (UL 489-1991) (IEEE Std C37.100-1992)

2.1.32 molded-case switch: A device designed to open and close a circuit by nonautomatic means, assembled as an integral unit in a supportive and enclosed housing of insulating material. (NEMA AB 1-1993) (UL 1087-1993) *Syn:* Nonautomatic switch (deprecated).

2.1.33 overcurrent: Any current in excess of the rated current of equipment or the ampacity of a conductor. It may result from overload, short-circuit, or ground fault. (NFPA 70-1996) *See also:* **overload.**

2.1.34 overload: Operation of equipment in excess of normal, full-load rating, or a conductor in excess of rated ampacity that, when it persists for a sufficient length of time, would cause damage or dangerous overheating. A fault, such as a short-circuit or ground fault, is not an overload. (NFPA 70-1996) *See also:* **overcurrent.**

2.1.35 panelboard: A single panel or group of panel units designed for assembly in the form of a single panel; including buses, automatic overcurrent devices, and equipped with or without switches for the control of light, heat, or power circuits; designed to be placed in a cabinet or cutout box placed in or against a wall or partition and accessible only from the front. (NFPA 70-1996) *See also:* **switchboard.**

2.1.36 peak current: The maximum instantaneous current that flows in a circuit, designated Ip. (NEMA AB 1-1993)

2.1.37 peak let-through current: The highest current flowing in the circuit following the inception of the fault that the circuit breaker and the protected system must withstand, expressed as an instantaneous rather than an rms value.

2.1.38 pickup: The rms current at which a circuit breaker tripping function is initiated. (NEMA AB 1-1993)

2.1.39 prospective fault current: The current that would flow during a short-circuit if the circuit breaker and the wires used for its connection were replaced by a solid conductor of negligible impedance. (UL 489-1991) *See also:* **available short-circuit current.**

2.1.40 rated short-time withstand current: (A) The maximum rms total current that a circuit breaker can carry momentarily without electrical, thermal, or mechanical damage or permanent deformation. The current shall be the rms value, including the dc component, at the major peak of the maximum cycle as determined from the envelope of the current wave during a given test time interval. (IEEE Std C37.100-1992) **(B)** That value of current assigned by the manufacturer that the device can carry without damage to itself, under prescribed conditions. (NEMA AB 1-1993) *Syn:* **withstand rating; short-time rating.**

2.1.41 rating plug: An interchangeable module of an electronic trip unit that, together with the sensor, sets the current rating range of the circuit breaker. For example, a 1200 A frame may contain an 800 A sensor, fixing the maximum rating that can be configured for the unit at 800 A adjustable by the following kind of settings. By installing a 600 A rating plug, the adjustable rating is correspondingly 600 A multiplied by the long-time pickup adjustment [i.e., the long-time pickup may be adjusted to "0.9" and the ampere rating or setting is (0.9 × 600 A) = 540 A.]

2.1.42 release: A device, mechanically connected within the circuit breaker, that initiates the tripping function of a circuit-breaker. (NEMA AB 1-1993)

2.1.43 rms sensing: A term commonly used to indicate the sensing of root-mean-square (rms) value current rather than instantaneous or peak values, as by a circuit-breaker trip unit.

2.1.44 selective coordination: *See:* **coordination; selectivity.**

2.1.45 selectivity: A general term describing the interrelated performance of relays and breakers, and other protective devices; complete selectivity being obtained when a minimum amount of equipment is removed from service for isolation of a fault or other abnormality. (IEEE Std C37.100-1992) *See also:* **coordination.**

2.1.46 sensor: (as applied to a circuit-breaker with an electronic trip unit) A current sensing element such as a current transformer within a circuit-breaker frame. The sensor will have a current rating less than or equal to the frame size and will provide the sensing function for a specific group of current ratings within the frame size.

2.1.47 series rating: The interrupting rating of a tested combination of a line-side (main) overcurrent protective device and a load-side (branch) circuit-breaker in which the interrupting rating of the combination is greater than the interrupting rating of the branch circuit-breaker. The interrupting rating of the series combination does not exceed the interrupting rating of the main overcurrent protective device. *Syn:* series-connected rating. See: UL Recognized Component Directory, 1996 [B1].[2]

2.1.48 setting: (of a circuit breaker) The value of current and/or time at which an adjustable circuit-breaker is set to trip. (NFPA 70-1996)

2.1.49 short-circuit: An abnormal connection (including an arc) of relatively low impedance, whether made accidentally or intentionally, between two points of different potential. (IEEE Std C37.100-1992)

2.1.50 short-time current: The current carried by a device, an assembly, or a bus for a specified short-time interval. (IEEE Std C37.100-1992) *See also:* **rated short-time withstand current; short-time rating.**

2.1.51 short-time delay: An intentional time delay in the tripping of a circuit-breaker which is above the overload pickup setting.

2.1.52 short-time delay phase or ground trip element: A direct-acting trip device element that functions with a purposely delayed action (measured in milliseconds). (IEEE Std C37.100-1992)

[2]The numbers in brackets correspond to those of the bibliography in 2.4.

2.1.53 short-time pickup: The current at which the short-time delay function is initiated. (NEMA AB 1-1993)

2.1.54 short-time rating: A rating applied to a circuit-breaker that, for reason of system coordination, causes tripping of the circuit-breaker to be delayed beyond the time when tripping would be caused by an instantaneous element. See also: rated short-time withstand current, short-time current. (UL 489-1991)

2.1.55 shunt trip device: A trip mechanism energized by a source of voltage that may be derived either from the main circuit or from an independent source. (UL 489-1991)

2.1.56 stored-energy operation: Operation by means of energy stored in the mechanism itself prior to the completion of the operation and sufficient to complete it under predetermined conditions. (IEEE Std C37.100-1992)

2.1.57 switchboard: A large single-panel, frame, or assembly of panels on which are mounted, on the face or back or both, switches, overcurrent and other protective devices, busses, and usually instruments. Switchboards are generally accessible from the rear as well as from the front and are not intended to be installed in cabinets. (NFPA 70-1996) *See also:* **panelboard.**

2.1.58 switchgear: A general term covering switching and interrupting devices and their combination with associated control, metering, protective and regulating devices; also assemblies of these devices with associated interconnections, accessories, enclosures, and supporting structures used primarily in connection with the generation, transmission, distribution, and conversion of electric power. (IEEE Std C37.100-1992)

2.1.59 transient recovery voltage (TRV): The voltage transient that occurs across the terminals of a pole of a switching device upon interruption of the current. (IEEE Std C37.100-1992)

2.1.60 tripping: The opening of a circuit breaker by actuation of the release mechanism. (NEMA AB 1-1993)

2.1.61 trip unit: A self-contained portion of a circuit-breaker that actuates the mechanism that opens the circuit-breaker contacts automatically. (NEMA AB 3-1991)

NOTE—The terms *trip device* and *tripping device* are used in literature as alternate terms for *trip unit.*

2.1.62 undervoltage trip device: A trip mechanism that causes a circuit-breaker to open automatically if the voltage across the terminals of the trip coil falls below a predetermined value. *Syn:* undervoltage release. (UL 489-1991)

2.1.63 withstand current: *See:* **rated short-time withstand current.**

2.1.64 withstand rating: *See:* **rated short-time withstand current.**

2.1.65 zone selective interlocking: A function provided for rapid clearing while retaining coordination. The function is a communication interconnection between the electronic trip units of two or more circuit-breakers connected in series on multiple levels. By means of intercommunication between the short-time delay and/or ground fault elements, the one nearest the fault trips with minimum time delay while signaling the supply-side circuit-breaker(s) to delay for a predetermined period. *Syn:* zone interlocking, selective interlocking.

2.2 Acronyms and abbreviations

The following acronyms and abbreviations are commonly used in marking and designating circuit-breaker ratings.

2.2.1 40C: Designates a circuit-breaker that is acceptable for use in ambient temperatures up to 40 °C.

2.2.2 AIR: Amperes Interrupting Rating. Shortened term marked on some small circuit-breakers with the interrupting rating.

2.2.3 CTL: A Class CTL circuit-breaker, because of its size or configuration in conjunction with the physical means provided in Class CTL panelboards, prevents more circuit-breaker poles from being installed than the number for which the assembly is designed and rated. A Class CTL panelboard is a circuit-limited lighting and appliance panelboard as referenced in the NEC, Sections 384-14 to 384-16. Both "half sized" and "full sized" circuit-breakers may be installed.

2.2.4 HACR: Heating, Air Conditioning, and Refrigeration. Designates compliance with the special requirements of NEC Section 430-53(c)(3) as "listed for group installation" for use with heating, air conditioning and refrigeration equipment. A circuit-breaker with this marking is suitable only for use with equipment marked to indicate that an HACR circuit-breaker is acceptable.

2.2.5 HID: High-Intensity Discharge. Indicates construction suitable for switching high-intensity discharge lighting loads.

2.2.6 SWD: Switching Duty. Designates compliance with requirements for circuit-breakers used as switches on fluorescent lighting circuits as indicated in NEC Section 240-83(d).

2.3 References

This chapter shall be used in conjunction with the following publications:

IEEE Std 100-1996, IEEE Standard Dictionary of Electrical and Electronics Terms.

IEEE Std C37.100-1992, IEEE Standard Definitions for Power Switchgear (ANSI).[3]

NEMA AB 1-1993, Molded Case Circuit Breakers and Molded Case Switches (DoD).[4]

NEMA AB 3-1991, Molded Case Circuit Breakers and Their Application.

NFPA 70-1996, National Electrical Code® (NEC®).[5]

UL 489-1991, Molded-Case Circuit Breakers and Circuit-Breaker Enclosures (DoD).[6]

UL 1087-1993, Molded-Case Switches.

UL Molded-Case Circuit Breaker Marking Guide, 1988.

2.4 Bibliography

Additional information may be found in the following source:

[B1] UL Recognized Component Directory, 1996.

[3]IEEE publications are available from the Institute of Electrical and Electronics Engineers, 445 Hoes Lane, P.O. Box 1331, Piscataway, NJ 08855-1331, USA.

[4]NEMA publications can be obtained from the National Electrical Manufacturers Association, 1300 N. 17th Street, Suite 1847, Rosslyn, VA 22209, USA.

[5]The NEC is available from Publication Sales, National Fire Protection Association, 1 Batterymarch Park, P.O. Box 9101, Quincy, MA 02269-9101, USA. It is also available from the Institute of Electrical and Electronics Engineers, 445 Hoes Lane, P.O. Box 1331, Piscataway, NJ 08855-1331, USA.

[6]UL standards are available from Global Engineering, 1990 M Street NW, Suite 400, Washington, DC, 20036, USA.

Chapter 3
Rating and testing

3.1 Relevance of rating and testing

This chapter provides information on rating and testing of circuit breakers that will be helpful to an electrical power systems engineer or specifying engineer in choosing the type of low-voltage circuit breaker to specify for an application. Ratings assigned to circuit breakers are determined by the testing done to prove their design capabilities. Therefore, a discussion of the testing requirements of the different classes of circuit breakers is fundamental to understanding their capabilities.

Other influences such as the National Electrical Code® (NEC®) (NFPA 70-1993),[1] industry practices, and environmental considerations can also affect circuit breaker choices. These are also discussed in areas where they relate to testing requirements. Application considerations such as choosing the type of trip unit, the type of time-current-characteristic, and the continuous current and interrupting ratings to use to satisfy coordination requirements are the subjects of other chapters of this book.

3.2 The ideal circuit breaker

The ideal circuit breaker (not realizable) would have no internal impedance and would carry current with no voltage drop when closed, and therefore would not produce any heat during operation. It would interrupt any overcurrent it was called upon to interrupt without experiencing contact erosion and structurally it would be able to withstand the pressure and heat of interruption and the magnetic forces produced by the flow of fault current through it. Its connectors would be firmly attached to the circuit breaker terminals and would hold the external circuit conductors in place regardless of the amount of fault current flow. It would provide perfect electrical isolation of the load from the normal system voltage when open and it would be able to provide these functions indefinitely regardless of the environment.

3.3 The practical circuit breaker

Practical circuit breakers do not, of course, perform exactly in the fashion described above. They have internal impedance and therefore they develop a small voltage drop across them while carrying current. The product of the resistive voltage drop and the current is a measure of power loss, which is manifested as heat in the circuit breaker during normal operation. The consequence is that practical circuit breakers warm up during operation.

When properly applied within their ratings and when they are in good operating condition, practical circuit breakers can interrupt any overcurrent that occurs in the circuit in which they

---

[1] Information on references can be found in 3.43.

are applied. They can do this without undue contact erosion and can withstand the pressure and heat of interruption as well as the magnetic forces produced by the flow of fault current. They will not damage the operating mechanism, trip unit, or supporting frame. Since their contacts do experience some erosion, they can only perform a limited number of switching and interrupting operations without maintenance where maintenance is possible. Circuit breakers that cannot be maintained should be replaced whenever wear and tear reaches certain limits. Properly sized and properly torqued connectors attached to the circuit breaker terminals can hold properly constructed conductors firmly in place in service.

Practical circuit breakers can provide adequate disconnecting capability to isolate the load from recovery voltage transients, surges, and normal voltage once they are opened and they can provide adequate service for an economically acceptable period of time.

3.4 Basic circuit breaker selection criteria

The selection of any circuit breaker, for any given duty, is ultimately based on an assessment of its ability to perform the following basic functions:

a) To carry the required full-load current without overheating (i.e., it should have the correct current rating),

b) To switch and isolate or disconnect the load from the source at the given system voltage (i.e., it should have the correct voltage rating),

c) To interrupt any possible abnormally high operating current or short-circuit current likely to be encountered during operation (i.e., it should have the correct interrupting rating), and

d) To be able to perform these functions over an acceptably long period of time under the operating and environmental conditions that will actually prevail in the application (i.e., it should have the correct mounting provisions, enclosure, and accessories for operation in the environment in which it is to be applied).

The degree to which a circuit breaker can satisfy these requirements is a measure of its applicability for a function. A circuit breaker's rating indicates these capabilities to the user because rating is established by proof testing. Hence, an understanding of how a circuit breaker of any given type is tested will give insight to its applicability for any function.

3.5 The role of standards

The primary vehicle for ensuring commonality in performance among circuit breakers of the same rating produced by different manufacturers is a product standard. Standards represent the consensus of manufacturers about what a given product should be able to do as a *minimum*. Standards establish the design tests that each manufacturer must perform and pass in order to claim a rating and to be in compliance with that standard. Some standards include requirements for periodic follow-up testing which, in effect, continues to sample the capabilities of newly manufactured circuit breakers. This assures that they maintain the capabilities of their product ratings. Standards provisions also provide for monitoring the quality of the materials used in the construction of circuit breakers and the quality of the workmanship in the manufacturing process.

As stated previously, standards requirements for the different classes of circuit breakers establish a basis for *minimum* performance. Circuit breakers may prove by test to perform better than their product ratings indicate, but they can never be permitted to perform worse. The user, however, may never assume that a circuit breaker can perform better than its rating indicates and should realize that there are manufacturing variations among mass-produced products. The levels of performance required by the standards for the minimum acceptable performance of different classes of circuit breakers will be the primary references in the discussion that follows.

3.6 The role of safety and industry codes

Safety and industry code requirements have evolved over the years. These codes reflect experience in actual application over the years. The Occupational Safety and Health Act (OSHA) [B12][2] is the primary legal safety code in the U.S. and the NEC is the primary safety code for construction. The NEC is often made part of state and local law, but to be effective, it must be accepted by the authority having jurisdiction. However, it can be ignored by local bodies and it can be modified. Other safety standards are the National Electrical Safety Code® (NESC®) (Accredited Standards Committee C2-1997) [B1] sponsored by the IEEE; NFPA 70B-1994, Electrical Equipment Maintenance [B10]; and NFPA 70E-1995, Electrical Safety Requirements for Employee Workplaces [B11]. Industry practices vary between industries and may vary from location to location in the same industry. The NEC and other industry codes for application may either make use of already available circuit breaker product features or they may influence the development of them. The feature of the sealable cover over the trip adjusting mechanism of a circuit breaker described later indicates how code provisions can affect product design and installation economics.

3.7 Comparison of testing requirements

In most cases, meaningful one-to-one comparison of test procedures will not be possible. It will be seen that tests of different types of circuit breakers differ and without the exact same testing, comparisons of the severity of tests can only be made subjectively. There are also some local differences in interpretation of the requirements of the NEC and other specific industry safety codes that make them also somewhat subjective. Only consideration and evaluation of the whole set of application requirements will permit a confident selection of a type of circuit breaker to be made. This chapter addresses many of the details that should be evaluated.

[2]The numbers in brackets correspond to those of the bibliography in 3.44.

3.8 Circuit breaker classes and types

For low-voltage circuit protection in the U.S., circuit breaker designs and tests are based on the requirements of three standards organizations—the American National Standards Institute (ANSI), Underwriters Laboratories (UL), and the National Electrical Manufacturers Association (NEMA). The two classifications of circuit breakers these organizations define are as follows:

— Molded-case circuit breaker class
— Low-voltage power circuit breaker class

Three types of circuit breakers are based on the two classifications above. The classifications themselves lend their names to the first two of the three types, while the third type, derived from the molded-case circuit breaker class, is known as an insulated-case circuit breaker. The three types of circuit breakers are as follows:

— Molded-case circuit breakers (MCCBs)
— Low-voltage power circuit breakers (LVPCBs)
— Insulated-case circuit breakers (ICCBs)

The following are some of the salient features of these types of circuit breakers.

MCCBs, as a class, are those tested and rated according to UL 489-1991 and whose current-carrying parts, mechanisms, and trip devices are completely contained within a molded case of insulating material. MCCBs are available in the widest range of sizes, from the smallest (15 A or less) to the largest (6000 A), and with various interrupting ratings for each frame size. They are characterized generally by fast interruption short-circuit elements. With electronic trip units they can have limited short-delay and ground-fault sensing capability. Virtually all MCCBs interrupt fast enough to limit the amount of prospective fault current let-through and some are fast enough and limiting enough to be identified as current-limiting circuit breakers. MCCBs are not designed to be field maintainable.

ICCBs are also rated and tested according to UL 489-1991. However, they utilize characteristics of design from both the power and molded-case classes. They are of the larger frame sizes, fast in interruption, but normally not fast enough to qualify as current-limiting circuit breakers. ICCBs also utilize electronic trip units and can have short-time ratings and ground-fault current sensing. They utilize stored energy operating mechanisms similar to those designed for LVPCBs and their design is such that they are partially field maintainable.

LVPCBs are rated and tested to satisfy ANSI C37 standard requirements and are used primarily in drawout switchgear. They are generally characterized as being the largest in physical size. They have short-time ratings, but they are not fast enough in interruption to qualify as current-limiting. LVPCBs are designed to be maintainable in the field.

The ANSI C37 series of standards and UL 489-1991 were jointly developed by IEEE and NEMA and apply to LVPCBs and ICCBs/MCCBs, respectively.

3.9 Generalized application considerations

Relative to the physical details of design and application, MCCBs are most often applied fixed mounted; however, drawout mechanisms have been designed for some of the largest ones and plug-in mechanisms have been designed for some of the smaller ones. Larger MCCBs are designed to utilize interchangeable trip units. They can be either thermal-magnetic trip units using bimetallic overload elements and electromagnetic overcurrent trips or they can be electronic trip units incorporating electrical analog or digital logic circuits to calculate current levels and initiate tripping functions. They usually derive operating power from current sensors mounted in the circuit breakers themselves.

ICCBs are used primarily in fixed mounted switchboards, but because of their size and weight, they are frequently mounted utilizing drawout mechanisms. ICCBs are designed primarily to utilize interchangeable state-of-the-art electronic trip units.

While considering the merits of maintainability in circuit breakers, it is well to remember that although MCCBs and ICCBs are not designed to be field maintainable, they are designed to have relatively high unmaintained endurance capabilities that should be evaluated in any application.

LVPCBs are always used in enclosures and because of their large size and weight are essentially always applied in drawout construction. Fixed mounting is an option rarely used. Drawout mechanisms not only minimize the work required in circuit breaker installation and maintenance, but they facilitate rapid changeout, which minimizes system down-time. This is especially important when the circuit breaker to be changed out is a main and a building or plant shutdown is necessary to change it. Drawout mechanisms also facilitate the performance of regular field inspection and maintenance services.

LVPCBs can utilize a variety of trip units including the latest versions of electronic trip units that are used in large MCCBs and ICCBs as well. Significant differences in the total operating times of different circuit breakers using either different types of trip units or the same type of electronic trip unit are accounted for in specific time-current curves (TCCs) published for the different circuit breaker and trip unit combinations.

3.10 References on rating and application

The latest versions of standards should always be referenced for circuit breaker information. Preferred ratings and application recommendations for LVPCBs are given in ANSI C37.16-1988. Standards for low-voltage ac power circuit breakers used in enclosures are given in IEEE Std C37.13-1990, and test procedures for LVPCBs used in enclosures are given in ANSI C37.50-1989. Application factors are discussed in the IEEE Std C37.20 series. LVPCBs are generally UL Listed and can be UL Labeled. Ratings and test procedures for MCCBs and ICCBs are found in UL 489-1991. Other IEEE Color Books complement this book, offering a very complete and comprehensive source of application information for all three types of circuit breakers.

3.11 Endurance considerations

Tables 3-1 through 3-3 summarize some specific mechanical and electrical endurance test parameters. The information in these tables is taken from ANSI C37.50-1989 and UL 489-1991. These standards do not lend themselves to one-to-one comparison. The tests or test conditions are different. It is necessary to make subjective judgments to determine which test procedure might be more severe than another.

Although choices may be made by subjective judgment of test procedures, an alternative method is to follow established practices that have a history of long-term successful performance. Fortunately, information on these practices is often available in the engineer's own facility and if it is not, references such as IEEE Std 493-1990 [B8], which discusses reliability in general and contains sets of data that can be of assistance in decision making, can be referred to. The comments in this chapter may be sufficient to resolve some questions on evaluation, but sometimes they may be sufficient only to indicate a direction for further, more detailed engineering investigation.

Other aspects of application, such as the size or weight of a circuit breaker, or its physical placement in a building or facility, or its maintenance requirements, and not factors directly pertaining to electrical rating and testing, can have as great an effect on the decision of what type of circuit breaker to choose in some cases.

Table 3-1—Circuit breaker endurance test parameters

Parameter	UL 489	ANSI C37.50
Enclosure	Smallest individual or open if mounted on metal plate	Minimum dimension test enclosure
Current	Rated	Rated
Voltage	100–105% Rated	100–105% Rated max (254 V, 508 V, or 635 V)
Power factor (ac)	0.75–0.80	0.85 max
Time constant (dc)	Not defined	Not covered
Frequency	48–62 Hz	48–72 Hz
Ambient	Not defined	Not defined
Ground fuse	30 A	30 A or 10 AWG (copper)

Source: [B13].

Table 3-2—Endurance test operations

Preferred frame sizes for MCCBs and ICCBs (A)	Number of cycles of operation			
	Per min (See NOTE 1)	With current	Without current	Total
50	6	6000	4000	10 000
100	6	6000	4000	10 000
125	5	4000 (See NOTE 2)	4000	8000 (See NOTE 2)
150	5	4000 (See NOTE 2)	4000	8000 (See NOTE 2)
200	5	4000	4000	8000
225	5	4000	4000	8000
400	4	1000	5000	6000
600	4	1000	5000	6000
800	1	500	3000	3500
1200	1	500	2000	2500
1600	1	500	2000	2500
2000	1	500	2000	2500
2500	1	500	2000	2500
3000	1 (See NOTE 3)	400	1100	1500
4000	1 (See NOTE 3)	400	1100	1500
5000	1 (See NOTE 3)	400	1100	1500
6000	1 (See NOTE 3)	400	1100	1500

NOTES

1—For circuit breakers rated more than 800 A, the endurance test may, at the option of the manufacturer, be conducted in groups of 100 load operations. No-load operations may be conducted between groups of load operations at the option of the manufacturer.

2—Where tests are required on samples having ratings of 100 A or less, 250 V or less, the number of operations shall be the same as for the 100 A frame.

3—Rate of operation: 1 cycle/min for first 10 operations; thereafter in groups of 5, at 1 cycle/min, with an interval between groups that is agreeable to all concerned.

Source: UL 489, October 24, 1990, Table 19.1.

3.12 Voltage rating considerations

MCCBs, ICCBs, and LVPCBs utilize the same standard nominal system voltages of 600 V, 480 V, and 240 V. However, MCCBs and ICCBs have additional rated voltages of 120 V, 120/240 V, 277 V, 347 V, 480Y/277 V, and 600Y/347 V, and LVPCBs have maximum voltages. A few comments on these considerations can be very instructive.

First, for MCCBs and ICCBs, the nominal voltage levels are maximum "not to exceed" voltages, while LVPCBs, on the other hand, have assigned "maximum" voltages of 254 V ac, 508 V ac, and 635 V ac. Second, the slash marks (/) between some of the voltage ratings have

Table 3-3—Circuit breaker mechanical/electrical endurance test comparison

Frame size (A)	Minimum operation rate cycles/hour		Number of operating cycles (see NOTE 5)							
			Between servicing (see NOTE 2)		With current		Without current		Total	
	UL 489	ANSI C37.50 (see NOTE 4)	UL 489	ANSI C37.50	UL 489	ANSI C37.50	UL 489	ANSI C37.50	UL 489	ANSI C37.50
100	360	—	See NOTE 3	—	6000	—	4000	—	10 000	—
150	300	—	See NOTE 3	—	4000	—	4000	—	8000	—
225	300	30	See NOTE 3	2500	4000	4000	4000	10 000	8000	14 000
400	240	—	See NOTE 3	—	1000	—	5000	—	6000	—
600	240	30	See NOTE 3	1750	1000	2800	5000	9700	6000	12 500
800	60	30	See NOTE 3	1750	500	2800	3000	9700	3500	12 500
1200	60	—	See NOTE 3	—	500	—	2000	—	2500	—
1600	60	30	See NOTE 3	500	500	800	2000	3200	2500	4000
2000	60	30	See NOTE 3	500	500	800	2000	3200	2500	4000
2500	60	—	See NOTE 3	—	500	—	2000	—	2500	—
3000	See NOTE 1	30	See NOTE 3	250	400	400	1100	1100	1500	1500
4000	See NOTE 1	30	See NOTE 3	250	400	400	1100	1100	1500	1500

NOTES
1—First 10 at 60/h; thereafter in groups of 5 at 60/h with interval between as acceptable to all.
2—Servicing means adjusting, cleaning, lubricating, and tightening.
3—Servicing not permitted.
4—May be conducted in groups of operations provided at least one group is not less than 120.
5—Tests on same breaker—sequence not defined.

Sources: ANSI C37.50-1989 and UL 489, December 30, 1991.

significance to circuit breaker design, application, and testing. NEC Section 240-83 refers to straight and slash voltage marking in a fine print note (FPN), which is quoted below:

> "A circuit breaker with a straight voltage marking, e.g., 240 V or 480 V, may be applied in a circuit in which the nominal voltage between any two conductors does not exceed the circuit breaker's voltage rating; except that a two-pole circuit breaker is not suitable for protecting a 3-phase corner-grounded delta circuit unless it is marked 1-phase/3-phase to indicate such suitability.
>
> A circuit breaker with a slash rating, e.g., 120/240 V or 480Y/277 V, may be applied in a circuit in which the nominal voltage to ground from any conductor does not exceed the lower of the two values of the circuit breaker's voltage rating and the nominal voltage between any two conductors does not exceed the higher value of the circuit breaker's voltage rating."

Voltage is a sensitive factor in interruption and voltage rating makes a difference in application. That difference can be seen most readily in the different interrupting ratings given for the same circuit breaker at different system voltages. The ratings are proven in testing.

Because of the benefits and associated limitations of different types of grounding systems, testing of slash voltage rated circuit breakers is different from the testing of straight voltage rated circuit breakers. IEEE Std 142-1991 [B5] discusses grounding considerations in detail. Slash voltage rated circuit breakers take advantage of the fact that the power system neutral is fixed at ground potential. Table 3-4 indicates the various tests required of MCCBs and ICCBs as a function of the different numbers of poles of the circuit breaker and the different voltage ratings to be applied. More detail on the tests and the circuits they are performed in follows. Table 3-4 emphasizes the significant difference in testing required for circuit breakers to be rated with a straight voltage and those to be rated with a slash voltage. The NEC FPN is an excellent reminder of the significance of voltage rating.

See Figure 3-1 for test circuit details.

For insulation testing, LVPCBs, MCCBs, and ICCBs are subjected to a 2200 V ac dielectric withstand voltage test, or an equivalent dc test, when new.

Table 3-4—Circuit breaker interrupting ability operations[a]

Poles	Frame rating	Circuit breaker ac voltage rating		Letters indicate diagram in Figure 21.1 of UL 489-1991						Total number of operations
				Operations on each pole			Common operations			
—	—	—	—	O	CO	O	O	CO	O	—
—	—	—	—	—	—	—	—	—	—	—
1	All	120, 240, 277, 347, 480, or 600		A	A	A	—	—	—	3
1	All	120/240 (tested in pairs)		—	—	—	B	B	B	3
2	All	240, 480, or 600		E	E	—	D	—	—	5
2	All	120/240		—	—	—	C	C	C	3
2	0–1200 A	480Y/277 or 600Y/347		L	L	—	C	—	—	5
2	All	1ϕ–3ϕ		E	E	—	H	—	—	5
3	0–1200 A	240, 480, or 600		G	G	—	F	—	—	7
3	1201–Up	240, 480, or 600		G	G	—	F	—	—	7
3	All	120/240		—	—	—	J	J	J	3
3	0–1200 A	480Y/277, 600Y/347		K	K	—	I	—	—	7
3	1201–Up	480Y/277, 600Y/347		K	K	—	I	—	—	7

Source: UL 489, December 30, 1991, Table 22.1.

[a]For the 125/250 V dc rating, the number of operations is the same as for the 120/240 V ac rating. For the 250 V dc rating, the number of operations is the same as for the 240 V ac rating.

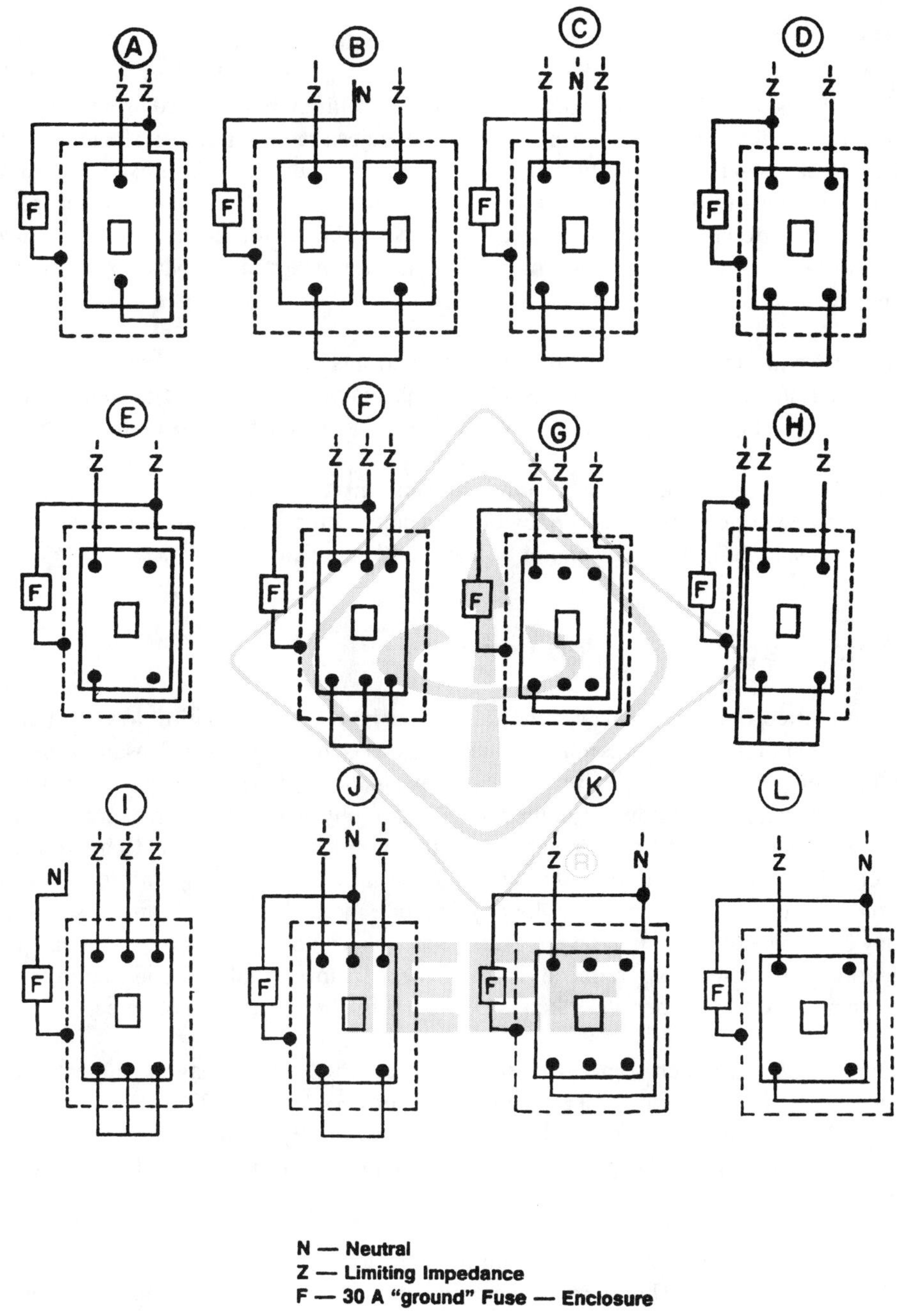

Figure 3-1—Interrupting ability test connection diagrams

Source: UL 489, October 9, 1991, Figure 22.1.

3.13 Frequency rating and considerations

In the U.S., MCCBs, ICCBs, and LVPCBs are all basically rated for 60 Hz operation. Whenever they are capable of being applied at other frequencies, they are marked to indicate those other frequencies. For operation at any frequency not specifically indicated as applicable, the manufacturer should be consulted. Sometimes no changes are required in performance specifications. Sometimes it may simply be necessary to recalibrate a circuit breaker trip unit for operation at the new frequency. Other times, for application at both lower or higher frequencies or for dc, a different trip unit may have to be used or the circuit breaker itself may have to be de-rated. Occasionally, application of a given circuit breaker might have to be absolutely prohibited at another frequency or dc.

With higher frequencies, the phenomena of eddy currents and skin effect have significance. They affect circuit breaker components (such as the primary current-carrying conductors or the iron cores of sensors) and/or accessory devices. The extent of their effects on operation determines whether a given circuit breaker can be de-rated for use or cannot be used at all at higher frequency. At lower frequency or dc, the method of current sensing may dictate whether a given circuit breaker can or cannot be used. Obviously, a circuit breaker utilizing transformer action for current sensing in its trip unit can not be used for dc applications.

3.14 Temperature considerations

Temperature affects circuit breaker operation in that below some limiting low temperature, mechanism operation will not be reliable due to possible freezing of condensation inside the circuit breaker, freezing of lubricants, and/or mechanical interference effects caused by changes in physical dimensions of components. Also, physical properties of materials may change. With extreme cold, some materials might tend toward brittleness. Temperature also affects circuit breaker operation in that above some limiting high temperature, mechanism operation will not be reliable because the physical or electrical strength limits of some materials may be reduced to marginal levels. Some materials can begin to melt and the useful life of insulation will be seriously reduced. Each of these factors should be considered in detail by the circuit breaker designer and taken into account by the application engineer.

Total temperatures to which some insulating materials in LVPCBs may be subjected to are listed in Table 3-5. This table is taken from IEEE Std C37.13-1990. These data are used by circuit breaker designers. Users of circuit breakers need only ensure that operation of the complete equipment will take place within the maximum and minimum limiting ambient temperatures.

The standard operating ambient temperature range for MCCBs and ICCBs is –5 °C to +40 °C (41 °F to 104 °F). The standard operating ambient temperature range for LVPCBs is also –5 °C to +40 °C (41 °F to 104 °F) but IEEE Std C37.20.1-1993 permits the temperature of the cooling air surrounding the enclosure of low-voltage switchgear to be within the range of –30 °C to +40 °C (–22 °F to +104 °F).

Table 3-5—Limits of temperature rise in circuit breaker components

	Limit of temperature rise over air surrounding enclosure (°C)	Limit of total temperature (°C)
Class 90 insulation	50	90
Class 105 insulation	65	105
Class 130 insulation	90	130
Class 166 insulation	115	155
Class 180 insulation	140	180
Class 220 insulation	180	220
Circuit breaker contacts, conducting joints, and other parts (except the following)	85	125
Fuse terminals	See NOTE 1	See NOTE 1
Series coils with over class 220 insulation or bare	See NOTE 1	See NOTE 1
Terminal connections (See NOTE 2)	55	95

NOTES
1—No specified limit except not to damage adjacent parts.
2—Terminal connection temperatures are based on connections to bus in low-voltage metal-enclosed switchgear. If connections are made to cables, recognition must be given to possible thermal limitations of the cable insulation and appropriate measures must be taken.

Source: IEEE Std C37.13-1990.

Since circuit breakers applied in temperatures outside their operating ranges may malfunction, the user should provide the necessary operating conditions required to keep the temperature surrounding the assemblies and circuit breakers in their safe operating range. Sometimes this will mean that space heaters are required inside an enclosure to prevent condensation. Other times, special enclosures will be required not only for temperature control, but also possibly for environmental protection against contamination by particulate matter, liquids, or gas vapors. When extreme cold or hot ambient conditions are possible, the addition of separate heating or air cooling systems is necessary.

Sometimes forced ventilation alone can be sufficient to transfer heat in a given ambient and the use of increased forced air flow rather than the addition of air cooling equipment can be of economic advantage. When ventilating fans are used to establish current-carrying capability, it should be remembered that the current-carrying capability so obtained is now dependent upon the operation of the fans, and fan performance may need monitoring.

Any necessary separate heating or air cooling equipment that is required for an application should be provided in the system design. It should be kept in mind that this equipment can be costly, can be heavy, and will occupy space. However, to obtain design life operation of switchgear and circuit breakers, ambient temperature limitations should be observed.

3.15 Enclosure considerations

All circuit breakers are fully rated for operation in free air at their listed maximum ambient temperature. All LVPCBs are applied in enclosures and all are also fully rated for operation in their design enclosures with maximum design ambient temperature. LVPCBs are therefore said to be 100% rated. Some MCCBs are also 100% rated.

The requirements for 100% rating of MCCBs are given in paragraph 19.25 of UL 489-1991. Parts of some of the pertinent details concerning 100% rating are as follows:

a) "A circuit breaker may be rated for continuous operation at 100% of its ampere rating if it (1) is of a frame size rated 250 A or more or a multi-pole type of any ampere rating and rated higher than 250 V..."

b) The candidate circuit breaker is tested for compliance with the 100% rating criteria in "an enclosure that is representative of the smallest enclosure with which the circuit breaker is likely to be used."

c) For compliance, "The temperatures of the insulating materials used in the circuit breaker shall not exceed the limits for the material involved.", and "The temperature rises (1) where connections are made to external bus bars, when bus bars are used; or (2) on a wiring terminal at a point to which the insulation of a wire is brought up... shall not exceed 60 °C (140 °F)..."

When the total terminal temperature in contact with insulation is 90 °C (194 °F), then 90° C (194 °F) rated conductor should be used. The minimum enclosure size for 100% rated application of MCCBs is indicated on the instruction sheet furnished with the circuit breaker and on the circuit breaker nameplate.

MCCBs that are not 100% rated are capable of operation in an enclosure at their rated maximum temperature at 80% of their free air current rating. Effective de-rating to 80% in application results from NEC rules stating that (1) the circuit must be wired for and the rating of the overload device shall not be less than the noncontinuous load plus 125% of the continuous load and that the circuit must be wired for 125% of full load ampacity and (2) the circuit breaker must be applied according to the ampacity of the circuit. NEC Section 220-10 (b) points out that the circuit breaker rating should not be less than the noncontinuous load plus 125% of the continuous load. Even with all continuous load, this results in the circuit breaker effectively being applied at no more than 80% of conductor ampacity or 80% of its own free air current rating.

NEC rules account for the application of 100% rated MCCBs by an exception to Section 220-10 (b). The 1993 NEC version of the exception states:

> "Where the assembly including the overcurrent devices protecting the feeder(s) are listed for operation at 100% of their rating, neither the ampere rating of the overcurrent device nor the ampacity of the feeder conductors shall be less than the sum of the continuous load plus the noncontinuous load."

Because circuit breakers are tested and rated to account for these factors, engineers should be aware of all aspects of the application and installation environment and specify circuit breakers tested to satisfy all of them.

3.16 Cable, wire, and conductor considerations

The type, size, and strand configuration of cable or wire conductor intended to be connected to a circuit breaker is an important application consideration and should satisfy at least two requirements. First, the safety code requirement for load-carrying ability states that feeder conductors shall have sufficient ampacity to supply the load served. Rules of the NEC should be followed to establish the minimum acceptable wire size. In order to satisfy voltage drop requirements, the conductors might be sized a little larger, but they can never be sized a little smaller; wires should always fit the connectors. Circuit breaker conductors also serve as conductive heat sinks for the circuit breaker. As heat sinks, the conductors should have a minimum cross-sectional conductor area equal to the cross-sectional conductor area of the 75 °C (167 °F) rated wire specified for that circuit breaker. They should also be properly constructed in stranding and compression and should have the correct insulation temperature rating.

Even though MCCBs are designed to operate in a 40 °C (104 °F) maximum ambient, the operating ambient temperature is not always 40 °C (104 °F). UL 489-1991 allows for the surface of conductor insulation to rise 35 °C (95 °F) above the normal ambient, which might be 25 °C (77 °F), a common normal room temperature, giving a total temperature of 60 °C (140 °F). With the maximum permissible rise of 50 °C (122 °F) on the terminals during temperature test, and again with a 25 °C (77 °F) ambient, the total temperature would be 75 °C (167 °F). These numbers are the basis for the 60 °C or 75 °C or 60 °C/75 °C conductor insulation ratings specified for the conductors used with MCCBs rated 125 A or less. For MCCBs rated greater than 125 A, the 75 °C conductor insulation rating is normal (called rated wire). When an MCCB is 100% rated, the maximum permissible terminal temperature rise during test is 65 °C (149 °F). When added to a 25 °C (77 °F) ambient, the total is 90 °C (194 °F). UL 489-1991 rules require that if the terminal temperature rise during 100% rating test exceeds 50 °C (122 °F), then the circuit breaker should be marked "For use with 90 °C (194 °F) wire and the wire size." The nameplate, in that case, would be marked accordingly.

Designers may want to use smaller, higher temperature rated wire. The application engineer should remember that when 90 °C (194 °F) insulated conductor is specified for a given ampacity, the cross-sectional area of the metal conductor inside the 90 °C (194 °F) rated insulation will generally be smaller than the cross-sectional area of the normal 75 °C (167 °F) rated conductor that was used to proof test the circuit breakers. If it is smaller, it will not be able to provide sufficient heat sinking capacity for normal circuit breaker performance and should not be used. Therefore, conductors connected to circuit breakers should have a cross-sectional area equal to that of the 75 °C (167 °F) rated conductor specified for the application.

Properly sized conductor, in addition to having sufficient ampacity, will also be stranded to satisfy the requirements of the circuit breaker terminals or connectors. It will have sufficient cross-section to adequately heat sink the circuit breaker, it will be insulated for the temperature conditions existing in all of the spaces through which it will pass, and it will be able to withstand fault interrupting forces and temperature changes without experiencing inordinate damage. Circuit breakers are tested with rated conductor to prove these capabilities. Therefore, rated wire size is sufficient for normal operation.

Many operating problems with circuit breakers start at the terminals. Therefore, all connectors used to connect conductors to circuit breakers at both the line side and the load side should be properly matched for size, material, and temperature rating and it should always be confirmed that they fit the circuit breaker terminals, that they are clean, and that they are torqued properly when installed. The connectors should be able to firmly hold the conductors in place against the forces that are imposed on them during short circuits. Standard interruption tests prove they can. It may be necessary to rope-tie conductors into place in some cases. When this is necessary, instruction sheets describe how it is to be done. The number of strands making up stranded conductor is an important factor in how well connectors can hold conductors against magnetic forces. See Table 8 of Chapter 9 of the NEC for a listing of normal conductor stranding to be used for circuit breaker wiring and see the notes below that table. The NEC requires stranding for conductors of size 8 AWG and larger.

Note that very flexible, finely stranded conductor, sometimes called welding cable, should not be used unless the connector is designed for it because the fine stranding is difficult or impossible to constrain under some standard terminal clamps. Sometimes the fine strands squeeze out from under the clamp, gradually loosening the connection. At the other extreme, conductors with fewer strands, but not compressed, can sometimes be tightly held with the first tightening, but as the conductor goes through heat expansion and contraction cycles or if the conductor is forced to move physically, the strands can rearrange themselves under the clamp, again resulting in a loose connection. Nothing could be worse for a circuit breaker. The higher resistance of loose conductors generates excess heat with rated current flow. It is not surprising, therefore, to hear comments to the effect that most circuit breaker problems are manifested at the terminals. The connectors specified by the manufacturer of the circuit breaker should be used because they are the ones used in proof testing. For more information on connectors and conductors, the interested reader is referred to the UL 486 series of standards on connectors ([B14] through [B18]).

All these factors affect circuit breaker operating temperature and are as important as ampacity. They are taken into account in the design and proof testing of circuit breakers.

As referred to previously, UL 489-1991 requires in paragraph 49.30 that

> "A circuit breaker, circuit breaker frame or interchangeable trip unit rated 125 A or less shall be marked as being suitable for 60 °C (140 °F), 75 °C (167 °F) only or 60/75 °C (140/167 °F) wire."

The indicated wire size and type should be used. Table 3-6 lists what are referred to as "rated wire sizes" for the various terminal currents or circuit currents. From the circuit breaker point of view, these wire sizes are a necessity. They are the sizes used in proof testing of circuit breaker designs.

Table 3-6—Terminal current and conductor size

Terminal current (A) (See NOTE 1)	Copper conductor			Aluminum or copper-clad aluminum conductor		
	No. of conductors	Size (AWG or kcmil)		No. of conductors	Size (AWG or kcmil)	
		60 °C	75 °C		60 °C	75 °C
15 or less	1	14	14	1	12	12
20	1	12	12	1	10	10
25	1	10	10	1	10	10
30	1	10	10	1	8	8
40	1	8	8	1	6	8
50	1	6	6	1	4	6
60	1	4	6	1	3	4
70	1	4	4	1	2	3
80	1	3	4	1	1	2
90	1	2	3	1	1/0	2
100	1	1	3	1	1/0	1
110	1	1	2	1	—	1/0
125	1	1/0	1	1	—	2/0
150	1	—	1/0	1	—	3/0
175	1	—	2/0	1	—	4/0
200	1	—	3/0	1	—	250
225	1	—	4/0	1	—	300
250	1	—	250	1	—	350
275	1	—	300	1	—	500
300	1	—	350	1	—	500
325	1	—	400	2	—	4/0
350	1	—	500	2	—	4/0

NOTES
1—For a terminal current other than indicated, the next higher rating is to be used (e.g., if rated 35 A, enter at 40 A).
2—Circuit breakers rated at more than 4000 A are to be considered as being bus- or cable-connected unless indicated otherwise in marking.
3—See 13.19 of UL 489, December 30, 1991.

Source: UL 489, December 30, 1991, Table 9.1.

Table 3-6—Terminal current and conductor size *(Continued)*

Terminal current (A) (See NOTE 1)	Copper conductor			Aluminum or copper-clad aluminum conductor		
	No. of conductors	Size (AWG or kcmil)		No. of conductors	Size (AWG or kcmil)	
		60 °C	75 °C		60 °C	75 °C
400	2	—	3/0	2	—	250
	1 (See NOTE 3)	—	500	1 (See NOTE 3)	—	750
450	2	—	4/0	2	—	300
500	2	—	250	2	—	350
550	2	—	300	2	—	500
600	2	—	350	2	—	500
700	2	—	500	3	—	350
800	3	—	300	3	—	400
1000	3	—	400	4 3	—	350 600
1200	4 3	—	350 500	4	—	500
1400	4	—	500	5	—	500
1600	5 4	—	400 600	5	—	600
2000	5	—	400 600	6	—	600
2500	8 7 6	—	400 500 600	8 7 9	—	600 750 500
3000	9 8 7	—	400 500 600	10 9 8	—	500 600 750
4000	12 11 10	—	400 500 600	13 12 11	—	500 600 750
5000 (See NOTE 2)	15 13 12	—	400 500 600	16 15 13	—	500 600 750
6000 (See NOTE 2)	18 16 15	—	400 500 600	19 18 16	—	500 600 750

NOTES
1—For a terminal current other than indicated, the next higher rating is to be used (e.g., if rated 35 A, enter at 40 A).
2—Circuit breakers rated at more than 4000 A are to be considered as being bus- or cable-connected unless indicated otherwise in marking.
3—See 13.19 of UL 489, December 30, 1991.

Source: UL 489, December 30, 1991, Table 9.1.

Maintenance personnel should remember that when a fault occurs, the conductors should be inspected for damage, and all damaged components should be repaired or replaced before reclosing the circuit breaker.

In summary, all conductors intended to be connected to circuit breakers should

a) Be large enough to carry the required maximum full-load current,
b) Be large enough to limit voltage drop to an acceptable application level,
c) Be large enough to withstand circuit breaker fault interruption let-through current,
d) Be small enough to fit into the circuit breaker connectors where they can be held tightly in place during a fault,
e) Be insulated for the rated system voltage,
f) Be of the correct stranding and construction to permit proper torquing, and
g) Have an insulation temperature rating and composition suitable for the total application.

Testing with wire confirms all these considerations.

3.17 De-rating for ambient temperature

MCCBs, ICCBs, and LVPCBs all should be de-rated in current-carrying capacity when operated in ambient temperatures above their rated maximum. The manufacturer of the circuit breaker should be consulted for the applicable de-rating information for a particular unit. There are formulae for calculation of current de-rating that are based on idealized simplifying assumptions and empirical factors. These formulae should be used with discretion.

For LVPCBs and ICCBs, the formula of 4.4.3.2 of IEEE Std C37.010-1979 can be used to determine a continuous-load current capability based on actual ambient temperature. That equation is as follows.

$$I_a = I_r \frac{(\theta_{max} - \theta_a)^{1/1.8}}{\theta_r} \tag{3-1}$$

where

I_a is the allowable de-rated current (A) (never to be more than two times I_r);
I_r is the rated continuous current (A);
θ_{max} is the allowable hottest spot total temperature = (θ_r + 40 °C);
θ_a is the actual ambient temperature (°C) expected;
θ_r is the allowable hottest spot temperature rise (°C) at rated current.

IEEE Std C37.010-1979 points out that the exponent generally has a range of 1/1.6 to 1/2.0 and that 1/1.8 is a compromise suitable for that application guide. The specifying engineer should always refer to the manufacturer for the best possible guidance in de-rating. Care should always be taken to ensure that the net result obtained is indeed that which was anticipated.

For MCCBs utilizing bimetallic overload trips, it is best to consult the manufacturer's temperature de-rating tables for the particular circuit breaker of interest because different bimetallic pairs operate at different temperatures. Obviously, trip mechanism designs and calibration methods can vary. However, the following general rule can be used to make rough estimates of expected thermal performance capability. Assuming temperature rise proportional to current squared, and taking the ratio of a known condition to an unknown condition, de-rated current can be solved as follows:

$$I_2 = I_1\sqrt{(T_2 - A_2)/(T_1 - A_1)} \tag{3-2}$$

where

- T_1 is the circuit breaker bimetallic element temperature, or the total terminal temperature for electronically tripped circuit breakers (°C) at rated current of I_1 amperes with rated ambient temperature A_1 °C;
- T_2 is assumed to remain approximately the same as T_1, not being too much affected by the practical difference in ambient temperatures;
- A_2 is the new ambient temperature (°C);
- I_2 is an estimate of the current (A) the circuit breaker is likely to be able to carry in an ambient temperature of A_2 °C.

This equation does not take into account any built-in compensation and as suggested can even be used to estimate thermal de-rating of electronically tripped circuit breakers by using the total terminal temperature for T_1 and T_2. For example, assuming a 50 °C (122 °F) terminal rise over a 40 °C (104 °F) ambient for a total temperature of 90 °C (194 °F) gives $T_1 = T_2 =$ 90 °C (194 °F) for rough estimating. The engineer should know the internal temperatures more accurately to reproduce manufacturer's de-rating data, but in the absence of manufacturer's data, this equation gives some guidance.

It should be remembered that the properties of the materials used in the construction of circuit-breaker components determine the maximum limiting temperature allowable for any given circuit breaker and they therefore also determine the amount of de-rating necessary for any given overtemperature condition. The properties of materials used in different circuit breakers can be different even when circuit breakers are similar in rating.

3.18 Circuit-breaker humidity limitations

The effect of humidity on any circuit breaker is a function of temperature. NEMA AB 1-1993 sets an operating limit on relative humidity in clean air at a level of not more than 50% at a maximum temperature of 40 °C. However, it recognizes that a higher level of as much as 90% relative humidity at a lower temperature of 20 °C could be satisfactory as long as consideration is given to the fact that moderate condensation is possible.

The detrimental effects of condensation are multiplied when water-soluble contaminants can also be present inside an enclosure. NEMA standardized enclosure types are available for various application conditions. When condensation is known to be possible in the application area, the circuit-breaker enclosures should at least be equipped with space heaters intended to prevent internal condensation by heating the air a small amount and allowing gravity to keep the internal air in motion.

3.19 Circuit-breaker altitude limitations

The altitude of an installation is important since, as altitude increases, atmospheric pressure and air density decrease. The reduced insulation and heat transfer properties of less dense air require that circuit breakers be de-rated for voltage withstand and current-carrying capacity as a function of altitude, assuming the temperature remains constant. Current de-rating can be compensated for to some degree if the temperature at the higher altitude is lower, but voltage withstand capability is essentially unaffected by lower temperature.

MCCBs and ICCBs should be de-rated for voltage, current-carrying capacity, and sometimes interrupting capacity, when applied at more than 6000 ft (1830 m) above mean sea level.

LVPCBs should be de-rated when applied 6600 ft (2000 m) above mean sea level. Table 3-7, taken from IEEE Std C37.13-1990, lists the specific altitude rating correction factors for LVPCBs. The manufacturers of MCCBs and ICCBs should be consulted for specific information on de-rating for altitude.

Table 3-7—Altitude rating correction factors

Altitude		Rated continuous current	Rated voltage
(ft)	(m)		
6600 and below	2000 and below	1.00	1.00
8500	2600	0.99	0.95
13000	3900	0.96	0.80
NOTE—Values for intermediate altitudes may be derived by linear interpolation.			

Source: IEEE Std C37.13-1990, Table 4.

There are few test sites in the world where actual altitude testing can be performed, and simulated altitude test facilities are few and far between. Instead, test voltage levels adjusted for the normal altitude conditions of the manufacturing site are used to test insulation integrity and, if necessary, rules based on theory and confirmed by experience are applied to answer practical interruption performance questions. The theoretical guidelines established for this purpose have proven to be quite satisfactory in application.

3.20 Circuit-breaker ampere rating

MCCB ampere ratings are characterized by a very wide range starting with the smallest single-pole lighting circuit breakers and ending with ratings shared by the largest LVPCBs. ICCB ampere ratings tend to be toward the high end of the MCCB range. LVPCB ampere ratings overlap a wide range of both MCCB and ICCB ratings with smallest power circuit breaker ratings in the neighborhood of a few hundreds of amperes.

Within a circuit breaker frame size, whenever trip units are interchangeable, the trip unit chosen determines the ampere rating of a particular circuit breaker. The trip units may be magnetic-only, thermal-magnetic, or electronic. In recent years, specialized high-technology trip units have been designed in packages that make them suitable for application in a number of the larger frame size circuit breakers of all three types.

3.21 National Electrical Code considerations

NEC Article 240, titled "Overcurrent Protection," gives guidance to the application engineer and the circuit breaker designer. NEC Section 240-3 states "Conductors, other than flexible cords and fixture wires, shall be protected against overcurrent in accordance with their ampacities as specified in Section 310-15...". Circuit breaker tests with wire and bus prove that circuit breakers can protect conductors. Standard wire ampacities are therefore of interest to the systems engineer and the circuit breaker designer when deciding on ampere ratings. Since the circuit breaker should protect the conductors, the choice of circuit breaker and conductor are related. When circuit breakers have trip units that fit into a single large frame, the following NEC consideration can be important to the circuit breaker and trip unit choice.

The ampere rating of a circuit breaker is as described in NEC Section 240-6, which lists the standard ampere ratings to be considered and further states "The rating of an adjustable trip circuit breaker having external means for adjusting the long-time pickup (ampere rating or overload) setting shall be the maximum setting possible." This ruling could affect project economics were it not for exception (b) to the rule. Without the exception, the rule would require wiring to be for the maximum circuit breaker trip level instead of for the maximum load current.

The exception to the rule states, "Circuit breakers that have removable and sealable covers over the adjusting means, or are located behind bolted equipment enclosure doors, or are located behind locked doors accessible only to qualified personnel, shall be permitted to have ampere ratings equal to the adjusted (set) long-time pickup settings." This exception gives the circuit breaker design engineer and the power system design engineer some latitude to affect power system economics. A simple feature like the provision of a sealable cover for the circuit breaker trip unit will allow a larger frame circuit breaker to be applied on a smaller ampacity circuit with the correct overload ampere trip setting. This feature makes it possible to keep the conductor size proportional to the circuit ampacity, thereby reducing the cost of the circuit conductors required, and it makes possible realization of economic benefits in commonality in the type of circuit breakers used and in the stocking of spares. Circuit breaker testing with different trip units make this possible.

NEC rules, of course, apply to all types of circuit breakers and consistent with the primary purpose of these rules and the exception (i.e., to protect the wire or bus and to do so safely), all circuit breakers are tested with rated wire or bus. This safety code provision recognizes the efficacy of circuit breaker test methods and offers the benefits of this demonstrated circuit breaker capability to users.

3.22 Preferred current ratings

The preferred frame sizes for MCCBs and ICCBs are listed in the first column of Table 3-2. Frame sizes for LVPCBs are listed in column 5 of Tables 3-8 and 3-9. It is from these lists that circuit breaker frame sizes are chosen. Table 3-10 indicates preferred ratings for LVPCBs that are integrally fused and utilize instantaneous direct-acting phase trip elements.

Table 3-8—Current ratings for low-voltage ac power circuit breakers with instantaneous direct-acting phase trip elements

Line no.	System nominal voltage (V)	Rated maximum voltage (V)	Insulation (dielectric) withstand (V)	3-phase short-circuit current rating (symmetrical A)[a]	Frame size (A)	Range of trip-device current ratings (A)[b]
	Col 1	Col 2	Col 3	Col 4	Col 5	Col 6
1	600	635	2200	14 000	225	40–225
2	600	635	2200	22 000	600	40–600
3	600	635	2200	22 000	800	100–800
4	600	635	2200	42 000	1600	200–1600
5	600	635	2200	42 000	2000	200–2000
6	600	635	2200	65 000	3000	2000–3000
7	600	635	2200	65 000	3200	2000–3200
8	600	635	2200	85 000	4000	4000
9	480	508	2200	22 000	225	40–225
10	480	508	2200	30 000	600	100–600
11	480	508	2200	30 000	800	100–800
12	480	508	2200	50 000	1600	400–1600
13	480	508	2200	50 000	2000	400–2000
14	480	508	2200	65 000	3000	2000–3000
15	480	508	2200	65 000	3200	2000–3200
16	480	508	2200	85 000	4000	4000
17	240	254	2200	25 000	225	40–225
18	240	254	2200	42 000	600	150–600
19	240	254	2200	42 000	800	150–800

Table 3-8—Current ratings for low-voltage ac power circuit breakers with instantaneous direct-acting phase trip elements *(Continued)*

Line no.	System nominal voltage (V)	Rated maximum voltage (V)	Insulation (dielectric) withstand (V)	3-phase short-circuit current rating (symmetrical A)[a]	Frame size (A)	Range of trip-device current ratings (A)[b]
	Col 1	Col 2	Col 3	Col 4	Col 5	Col 6
20	240	254	2200	65 000	1600	600–1600
21	240	254	2200	65 000	2000	600–2000
22	240	254	2200	85 000	3000	2000–3000
23	240	254	2200	85 000	3200	2000–3200
24	240	254	2200	130 000	4000	4000

[a]Ratings in this column are rms symmetrical values for single-phase (2-pole) circuit breakers and three-phase average rms symmetrical values of three-phase (3-pole) circuit breakers. When applied on systems where rated maximum voltage may appear across a single pole, the short-circuit current ratings are 87% of these values. See 5.6 of IEEE Std C37.13-1981.
[b]For preferred trip-device current ratings, see Table 22 of ANSI C37.16-1988. Note that the continuous-current-carrying capability of some circuit-breaker trip-device combinations may be higher than the trip-device current rating. See 10.1.3 of IEEE Std C37.13-1981.

Source: ANSI C37.16-1988, Table 1.

Table 3-11 lists current values taken from NEC Section 240-6.

Circuit breaker manufacturers design trip units for the various preferred ampere levels spanned by different frame sizes. All circuit breakers are tested with the various trip units installed to demonstrate both their time-current tripping characteristics and the circuit breaker's interrupting capability with that trip unit installed. The ability of a circuit breaker to protect rated cable or bus is demonstrated simultaneously since the test circuit is made up with specified size bus or wire of "rated wire size." See the individual manufacturer's literature for available trip unit ampere ratings.

Table 3-12 lists the preferred trip-device current ratings or settings for LVPCBs.

Table 3-9—Preferred current ratings for LVPCBs without instantaneous direct-acting phase trip elements (short-time-delay element or remote relay)

Line no.	System nominal voltage (V)	Rated maximum voltage (V)	Insulation (dielectric) withstand (V)	Three-phase short-circuit current rating or short-time current rating (symmetrical A)[a, b]	Frame size (A)	Range of trip-device current ratings (A)[c] Setting of short-time-delay trip element		
						Minimum time band	Intermediate time band	Maximum time band
	Col 1	Col 2	Col 3	Col 4	Col 5	Col 6	Col 7	Col 8
1	600	635	2200	14 000	225	100–225	125–225	150–225
2	600	635	2200	22 000	600	175–600	200–600	250–600
3	600	635	2200	22 000	800	175–800	200–800	250–800
4	600	635	2200	42 000	1600	350–1600	400–1600	500–1600
5	600	635	2200	42 000	2000	350–2000	400–2000	500–2000
6	600	635	2200	65 000	3000	2000–3000	2000–3000	2000–3000
7	600	635	2200	65 000	3200	2000–3200	2000–3200	2000–3200
8	600	635	2200	85 000	4000	4000	4000	4000
9	480	508	2200	14 000	225	100–225	125–225	150–225
10	480	508	2200	22 000	600	175–600	200–600	250–600
11	480	508	2200	22 000	800	175–800	200–800	250–800
12	480	508	2200	42 000	1600	350–1600	400–1600	500–1600
13	480	508	2200	50 000	2000	350–2000	400–2000	500–2000
14	480	508	2200	65 000	3000	2000–3000	2000–3000	2000–3000

Table 3-9—Preferred current ratings for LVPCBs without instantaneous direct-acting phase trip elements (short-time-delay element or remote relay) *(Continued)*

Line no.	System nominal voltage (V)	Rated maximum voltage (V)	Insulation (dielectric) withstand (V)	Three-phase short-circuit current rating or short-time current rating (symmetrical A)[a, b]	Frame size (A)	Range of trip-device current ratings (A)[c] Setting of short-time-delay trip element		
						Minimum time band	Intermediate time band	Maximum time band
	Col 1	Col 2	Col 3	Col 4	Col 5	Col 6	Col 7	Col 8
15	480	508	2200	65 000	3200	2000–3200	2000–3200	2000–3200
16	480	508	2200	85 000	4000	4000	4000	4000
17	240	254	2200	14 000	225	100–225	125–225	150–225
18	240	254	2200	22 000	600	175–600	200–600	250–600
19	240	254	2200	22 000	800	175–800	200–800	250–800
20	240	254	2200	42 000	1600	350–1600	400–1600	500–1600
21	240	254	2200	50 000	2000	350–2000	400–2000	500–2000
22	240	254	2200	65 000	3000	2000–3000	2000–3000	2000–3000
23	240	254	2200	65 000	3200	2000–3200	2000–3200	2000–3200
24	240	254	2200	85 000	4000	4000	4000	4000

[a]Short-circuit current ratings for breakers without direct-acting trip devices, opened by a remote relay, are the same as those listed here.
[b]Ratings in this column are rms symmetrical values for single phase (2-pole) circuit breakers and three-phase average rms symmetrical values of three-phase (3-pole) circuit breakers. When applied on systems where rated maximum voltage may appear across a single pole, the short-circuit current ratings are 87% of these values. See 5.6 of IEEE Std C37.13-1981.
[c]For preferred trip-device current ratings, see Table 22 of ANSI C37.16-1988. Note that the continuous-current-capability of some circuit-breaker trip-device combinations may be higher than the trip-device current rating. See 10.1.3 of IEEE Std C37.13-1981.

Source: ANSI C37.16-1988, Table 2.

Table 3-10—Preferred ratings for LVPCBs integrally fused with instantaneous direct-acting phase trip elements

Line no.	Circuit-breaker frame size (A)[a]	Rated maximum voltage (V)[b]	Insulation (dielectric) withstand (V)	Three-phase short-circuit current rating (symmetrical A)[c]	Range of continuous-current rating (A)	
					Range of trip-device current ratings (A)[d]	Maximum fuse rating[e]
	Col 1	Col 2	Col 3	Col 4	Col 5	Col 6
1	600	600	2200	200 000	125–600	f
2	800	600	2200	200 000	125–800	f
3	1600	600	2200	200 000	200–1600	f

[a]Two circuit-breaker frame ratings are used for integrally fused circuit breakers. The continuous-current rating of the integrally fused circuit breaker is determined by the rating of either the direct-acting trip device or the current-limiting fuse applied to a particular circuit-breaker frame rating, whichever is smaller.
[b]Listed values are limited by the standard voltage rating of the fuse.
[c]Ratings in this column are rms symmetrical values for single-phase (2-pole) circuit breakers and three-phase average rms symmetrical values of three-phase (3-pole) circuit breakers. When applied on systems where rated maximum voltage may appear across a single pole, the short-circuit current ratings are 87% of these values. See 5.6 of IEEE Std C37.13-1981.
[d]For preferred trip-device current ratings, see Table 22 of ANSI C37.16-1988. Note that the continuous-current-carrying capability of some circuit-breaker trip-device combinations may be higher than the trip-device current rating. See 10.1.3 of IEEE Std C37.13-1981. Lower rated trip-device current ratings may be used when the fuse size is small or the available current is low, or both. Consult the manufacturer.
[e]Fuse current ratings may be 300 A, 400 A, 600 A, 800 A, 1000 A, 1200 A, 1600 A, 2000 A, 2500 A, and 3000 A. Fuses are of the current-limiting type.
[f]Values have not yet been determined; consult the manufacturer.

Source: ANSI C37.16-1988, Table 17.

Table 3-11—Standard ampere ratings for inverse-time circuit breakers

15	70	225	1000
20	80	250	1200
25	90	300	1600
30	100	350	2000
35	110	400	2500
40	125	500	3000
45	150	600	4000
50	175	700	5000
60	200	800	6000

Source: 1993 NEC, Section 240-6.

Table 3-12—Preferred trip device current ratings or settings for LVPCBs (in amperes)[a]

40	225	1000
50	250	1200
70	300	1600
90	350	2000
100	400	2500
125	500	3000/3200
150	600	4000
175	800	5000[b]
200		6000

NOTE—Where these exact ratings are not available in solid-state trip devices, they may be closely approximated by the pickup setting of the long-time-delay element.

[a]See Tables 1, 2, 8, and 17 of ANSI C37.16-1988 for ranges of trip-device current ratings by circuit-breaker frame size.
[b]These values are for dc circuit breakers only.

Source: ANSI C37.16-1988, Table 22.

3.23 Load effects

Continuous-current ratings required of circuit breakers can be affected by the characteristics of the loads being served. Harmonics in nonlinear loads and high, short-time inrush loads like tungsten filament lighting and motor and transformer startup can affect the circuit breaker contacts or trip circuit logic. These load effects are applicable to all circuit breakers both in their thermal effect on the circuit breaker and their effect on the functioning of trip units. Tungsten filament lamp rating, heating and air conditioning rating (HACR), and high-intensity-discharge (HID) rating for example, are specially tested load ratings.

3.24 The effect of nonlinear loads on circuit breakers

Although capacitor banks do not generate harmonic currents themselves, they are low-impedance sinks for power system harmonic currents. Regardless of where the harmonic currents originate, if they flow into a capacitor bank, they do so by flowing through the associated circuit breaker or fuse controlling the capacitor bank. Modern drives and controls utilizing electronic power switching devices for control generate harmonic currents that propagate throughout the power system as a function of the impedances of the various paths. In order to provide adequate thermal protection for capacitor banks, circuit breaker trip units should be able to respond adequately to rms power regardless of harmonic content. Bimetallic thermal elements respond to rms directly through the heat produced in them, and newer electronic trip units have been designed with sampling algorithms to provide rms current sensing in the presence of harmonics. For more information on capacitor bank applications in particular, refer to NEMA CP 1-1988.

Resistance welding applications can also generate harmonics. Rapid rises in current followed by sharp cutoffs constitute nonsinusoidal current waveform during a welding operation. The nominal load level is the average of the rms peaks and valleys over a period of time. The duty cycle of a process, even with sinusoidal waveform, can also affect the required circuit-breaker rating. Refer to the NEC for detailed safety code guidance for these applications. Load nonlinearity can affect circuit breakers through trip unit sensitivity to current peaks as well as through circuit breaker component heating. Test procedures do not presently include provisions to account for nonlinear load effects, but as such loads continue to proliferate, the effects of nonlinearity can be expected to become more noticeable on the power system and ultimately engineers can expect to be required to take them into account in testing procedures.

Many modern microprocessor-based signal processing trip units include a heating memory algorithm based on preceding current level history so they can take into account the heating effect of that current flow just as bimetallic elements and real utilization devices do. With these capabilities they can very adequately protect power circuits feeding nonlinear loads.

At this time there is no commonly accepted specification of a test waveform that microprocessor trip units should be able to sample and accurately quantify. However, circuit-breaker manufacturers have done extensive development testing to provide assurance that their digital sampling algorithms will work to the degree that they specify in their technical literature.

3.25 The effect of high inrush loads

If loads are cycled on and off or up and down in level or if motor plugging[3] is involved in an application, the current rating and type of circuit-breaker trip unit should be chosen carefully. It may be necessary to consult the manufacturer of any type of circuit breaker for guidance under these conditions. Large inrush current with potentially high offset peaks and longer duration high continuous-current flows accompanies these operations, making them unlike normal motor starting duty, and these operations may occur more frequently in a process. The response of the circuit-breaker trip unit to these higher currents and offsets and their heating effect on the circuit breaker itself should be evaluated. Such effects are considered in circuit-breaker overload testing.

3.26 Overload testing of circuit breakers

MCCBs, ICCBs, and LVPCBs are tested to interrupt overload current at 600% of rated current a given number of times. ICCB and MCCB minimum overload test current is 150 A. See Tables 3-13 and 3-14. Tables 3-15 and 3-16 summarize the overload testing parameters and indicate data for comparison.

The definitions of overcurrent and overload in NEC Article 100 can be helpful in understanding short-circuit interrupting and overload testing.

> *Overcurrent.* "Any current in excess of the rated current of equipment or the ampacity of a conductor. It may result from overload (see definition), short-circuit, or ground fault."

> *Overload.* "Operation of equipment in excess of normal, full-load rating, or of a conductor in excess of rated capacity which, when it persists for a sufficient length of time, would cause damage or overheating. A fault, such as a short circuit or ground fault, is therefore not an overload."

It is important not to confuse these terms.

[3]Motor plugging is the process of reversing the polarity of dc motors or reversing the phase-rotation of ac motors and applying power to rapidly stop or change the speed of the motor's coupled load.

Table 3-13—Overload test operations for MCCBs and ICCBs[a]

Number of operations			
Frame size (A)	Close and open manually [b, c, d]	Close manually, open automatically	No. of cycles of operation/min
50	35	15	6
100	35	15	6
125	60[e]	—[e]	5
150	50[e]	—[e]	5
200	50	—	5
225	50	—	5
400	50[c]	—	4[f]
600	50[c]	—	4[f]
800	50[c]	—	1[f]
1200	50[c]	—	1[f]
1600	50[c]	—	1[f]
2000	25[c]	—	1[f]
2500	25[c]	—	1[f]
3000	28[g]	—	1[g]
4000	28[g]	—	1[g]
5000	28[g]	—	1[g]
6000	28[g]	—	1[g]

[a]The operations may be performed by a machine simulating manual operation.
[b]If the test sample trips during manual operation, it is still considered manual operation.
[c]At the option of the manufacturer, the adjustable instantaneous response of a circuit breaker rated 400 A or more may be adjusted to less than the maximum position.
[d]The minimum closed time shall be one cycle, unless the sample trips.
[e]In the case of a multiple breaker without a common trip, rated at more than 100 A, 35 cycles of operations are to be made manually and 15 are to be made automatically as covered in 16.10 of UL 489-1991.
[f]Operations may be conducted in groups of 5, with 15 min maximum between groups.
[g]Three operations at 600% of current rating at the rate of 1 cycle/min, followed by 25 operations at 200% of current rating at the rate of 1 cycle/min may be conducted in groups of 5 with 15 min maximum between groups.

Source: UL 489, October 24, 1990, Table 16.1.

Table 3-14—Overload switching requirements for low-voltage ac power circuit breakers

Line no.	Circuit-breaker frame size (A)	No. of make-break operations
	Col 1	Col 2
1	225	50
2	600	50
3	800	50
4	1600	38
5	2000	38
6	3000	N/A
7	3200	N/A
8	4000	N/A

Source: ANSI C37.16-1988, Table 3.

3.27 Safety factor for current loading

Normally, for all circuit breakers, continuous current is not to exceed the continuous-current rating of the particular frame size or overload device. However, certain normal operations like the starting of motors and the energizing of transformers will result in short-term or transient overload conditions. In the case of motor starting, the symmetrical inrush current to be expected was on the order of 600% of rated full-load motor current for as much as several seconds during starting. Today, with high-efficiency motors being used more often, it is even higher. But, motor starting does not usually occur very frequently. Therefore, this condition does not generally affect the choice of frame size or trip unit ampere rating. However, if ambient temperature is unusually high, or if starting time is unusually long, or if starting inrush is unusually high, some additional investigation is appropriate. The manufacturer should be consulted for guidance in those special cases. Energization of transformers is accompanied by large surges of excitation current, but they are very short lived compared to motor starting and need be considered only as they might affect a magnetic or instantaneous trip. All of these considerations apply equally to all types of circuit breakers. Time-current curves for circuit breakers and trip units indicate their specific time and current limitations. Circuit-breaker and trip-unit tests confirm these relationships.

Table 3-15—Overload performance test parameters[a]

Frame size (A)	Min operation rate (cycle/h)		No. of operating cycles					
			Opening manually		Opening automatically		Total	
	UL 489	ANSI C37.50	UL 489	ANSI C37.50[b, c]	UL 489	ANSI C37.50	UL 489	ANSI C37.50
100	360	—	35	—	15	—	50	—
150	300	—	50	—	—	—	50	—
225	300	60	50	50	—	—	50	50
400	240[d]	—	50[a]	—	—	—	50	—
600	240[d]	60	50[a]	50	—	—	50	50
800	60[d]	60	50[a]	50	—	—	50	50
1200	60[d]	—	50[a]	—	—	—	50	—
1600	60[d]	—	50[a]	38	—	—	50	38
2000	60[d]	—	25[a]	38	—	—	25	38
2500	60[d]	—	25[a]	—	—	—	25	—
3000	60[d]	—	28[a, e]	N/A	—	—	28	N/A
4000	60[d]	—	28[a, e]	N/A	—	—	28	N/A

[a]At option of manufacturer, adjustable instantaneous may be set at less than maximum.
[b]Must carry current minimum 1 cycle.
[c]Circuit opened by separately operated shunt trip device.
[d]Operations may be in groups of 5 with 15 min max.
[e]Three operations at 600% followed by 25 at 200%.

Source: [B13].

Table 3-16—Overload test parameters

Parameter	UL 489	ANSI C37.50
Enclosure	Smallest individual	Min dimension test enclosure
Current (ac) (dc)	6 × rated 6 × rated	6 × rated
Voltage	100–105% rated	100–105% rated max (254 V, 508 V, or 635 V)
Power factor (ac)	0.45–0.50	0.5
Time constant (dc)	3 ms min	Not covered
Frequency	48–62 Hz	48–72 Hz
Ambient	Not defined	Not defined
Ground fuse	30 A	30 A fuse or 10 AWG (copper)

Source: [B13].

3.28 Forced-air cooling of LVPCBs

In the case of LVPCBs, forced-air cooling may be utilized as an option to obtain additional continuous-current-carrying capacity. Although a system might be intentionally designed to utilize forced-air cooling, it is most likely that this approach would be decided upon as a backup contingency only since the ability of the circuit breakers to carry the required load current in this mode depends upon the operation of the air-moving fans. The reliability of the fans becomes a most important factor in the overall reliability of the circuit-breaker system.

Obviously, testing is necessary to prove the current-carrying capability of a circuit breaker with forced-air cooling because modeling of air flow and heat transfer inside circuit breakers and enclosures is a very difficult process at best and unexpected failure of a system can have unacceptable consequences. If forced-air cooling appears to be required or of advantage, the engineer should see 10.1.3.1 of IEEE Std C37.13-1990 for a detailed discussion of the factors to be considered and discuss the matter further with the manufacturer.

3.29 Short-circuit interrupting rating

Short-circuit interrupting rating addresses the ability of a circuit breaker to interrupt the actual flow of fault current in a circuit having a given prospective fault-current level and to protect the conductors connected to the circuit breaker. Circuit-breaker interrupting rating is now given on a symmetrical rms ampere basis. At one time it was given in asymmetrical amperes. The symmetrical short-circuit interrupting rating of the circuit breaker takes into account the initial current offset due to circuit X/R ratio. The value of the standard X/R ratio is used in the test circuit and its effect is therefore included in the interrupting rating. For

LVPCBs, the implicit value of X/R is 6.6 (15% power factor). For MCCBs and ICCBs, the implicit value of X/R used in the test circuit varies with the short-circuit current rating of the circuit breaker, having a maximum value of 4.898 (20% power factor). See IEEE Std 399-1990 [B7] for a more detailed discussion of short-circuit studies.

One benefit of rating circuit breakers on a symmetrical current basis is that the symmetrical short-circuit interrupting ampere level required for an application can be calculated much easier. Ohm's Law or any other method of circuit analysis can be used. The proposed IEEE Violet Book [B9] will discuss different methods for calculating short-circuit currents. Circuit-breaker evaluation is then made on the basis of the application circuit X/R ratio as compared to the test circuit X/R ratio. Tables of X/R facilitate the evaluation.

Short-circuit testing is done to ensure that a given frame size circuit breaker itself is capable of withstanding the heat and forces of a short-circuit interruption and that it can protect the conductors connected to it. The standards require proof testing of a circuit breaker's ability to interrupt bolted faults. When a circuit breaker of a given ampere rating is chosen for an application, its interrupting rating is chosen to be equal to or greater than the calculated short-circuit symmetrical current of the supply system at the point where the circuit breaker is to be connected to the supply system. The current calculated for this condition is called the prospective fault current. Actual fault current can never reach this level because there is always additional impedance added between the point of circuit breaker connection to the supply bus and the load side of the circuit breaker. The connections themselves and the circuit-breaker impedance are between those points.

The short-circuit interruption testing specified in standards takes this into account in different ways. For more discussion of the fault calculation process and the effect of fault impedance on the results, see IEEE Std 141-1993 [B4]. For details on the differences in testing, several standards should be reviewed. Generally, for LVPCB testing details, see ANSI C37.50-1989 and for MCCBs and ICCBs, see UL 489-1991.

The short-circuit current interrupting rating of a circuit breaker is that value of symmetrical short-circuit current that would flow in the circuit where the circuit breaker is to be connected for test. This is the test prospective fault current and is actually a measured value of current to ensure that the proper test conditions exist. The X/R ratio in the prospective test current circuit is set to the value specified for it in the applicable circuit-breaker standard. The circuit breaker is tested to prove that it is able to safely interrupt the fault current that actually flows from this circuit during the test.

The short-circuit current interrupting requirement for a circuit breaker to be applied in a practical system is called the prospective fault current for that system and is the value of symmetrical short-circuit current mathematically calculated for that system at the point of circuit-breaker application. The process of determining whether the circuit-breaker rating is sufficient to interrupt the prospective application circuit fault current is called circuit-breaker evaluation. More discussion of the circuit-breaker evaluation process follows.

3.30 Fault-current calculation considerations

Short-circuit application duty requirements for circuit breakers are calculated the same way for all types of circuit breakers. Since all circuit breakers are now rated on a symmetrical current basis, the initial step is an Ohm's Law solution of the symmetrical three-phase circuit. The short-circuit interrupting duty requirement for a circuit breaker is taken to be the short-circuit capacity of the power system at the point in the power system where the circuit breaker is to be connected. As discussed previously, this leads to conservative results since neither connection impedance, circuit-breaker impedance, fault-circuit impedance, nor arc impedance is taken into account in this calculation. Further, and better from the point of view of consistency, the results of the Ohm's Law system study calculations are essentially the same no matter which engineer does the calculation (provided commonly accepted power system data are used). Finally, any correction needed to account for fault-circuit power factor less than the value used for testing (or for X/R ratio greater than that used for testing) is applied. This is most often done by multiplying the fault current by a multiplying factor that is a function of the system X/R. A separate discussion of the effects of X/R ratio follows in this chapter. The proposed IEEE Violet Book [B9] will be an excellent reference for further study of the process of calculation.

3.31 Circuit-breaker interrupting ratings

Recognized current interrupting ratings for MCCBs and ICCBs are listed in Table 3-17. The preferred short-circuit interrupting ratings of LVPCBs are listed in Tables 3-8 and 3-9. Circuit breakers are designed with the goal of achieving one of these ratings.

Table 3-17—Current interrupting ratings for MCCBs and ICCBs—rms symmetrical or dc amperes

7500	25 000	65 000
10 000	30 000	85 000
14 000	35 000	100 000
18 000	42 000	125 000
20 000	50 000	150 000
22 000	—	200 000

Source: UL 489, October 9, 1991, Table 67.1.

MCCBs and ICCBs are tested in the prospective fault test circuit by connecting the circuit breaker on test in place of the shorting bus links used for test circuit calibration. The connections are made with lengths of rated wire or bus in accordance with UL 489-1991. The prospective current source or test laboratory source remains as set during calibration. Power factor values for the test circuit are as given in Table 3-18.

Table 3-18—Test circuit power factor for testing MCCBs and ICCBs

Test circuit (A)	Power factor
10 000 or less	0.45–0.50
10 001–20 000	0.25–0.30
over 20 000	0.15–0.20

Source: UL 489, October 9, 1991, Table 22.4.

LVPCBs can themselves be included in the prospective current test circuit when that circuit is being calibrated for testing. However, they usually are not. Shorting links are used to complete the test circuit for calibration and when the circuit is calibrated, the circuit breaker to be tested is connected into the prospective circuit to replace the short-circuiting links. Tests are then performed to satisfy one of the preferred short-circuit interrupting ratings given in Tables 3-8 through 3-10.

3.32 Single-pole fault interruption testing

Single-pole, maximum line-to-line voltage testing is done at the theoretical maximum single-phase fault-current level of 87% of maximum bolted three-phase fault current on all LVPCBs. Some MCCBs are tested similarly except at full-rated voltage that is equal to their maximum voltage. Other MCCBs are single-pole tested in a similar manner at the same full-rated voltage but at a reduced fault-current level. Tables 3-19 and 3-20 show the test current values used. Refer to IEEE Std 242-1986 [B6] for more discussion of single-pole considerations.

3.33 Circuit-breaker evaluation in standards for testing

Voltage, symmetrical short-circuit current magnitude, and circuit X/R ratio as seen from the point of circuit-breaker connection to the power system are the factors that should be known to evaluate a circuit breaker. If a circuit breaker does not have an ANSI or UL or other third-party certification based on a testing standard, then it may be very difficult or impossible to evaluate. Its method of testing should be known to determine its interrupting capability. Otherwise, it cannot be evaluated and should not be applied.

Present IEC practices and standards do not directly correspond to the practices and standards in use in North America for single-pole duty, thermal response, or grounding. This can make it very difficult at best to make comparison evaluations between domestic and IEC circuit-breaker capabilities. UL 489-1991 and ANSI C37.50-1989 short-circuit testing procedures and parameters are tabulated for reference in Tables 3-21 through 3-23 of this chapter. These tables show how the differences between procedures can complicate direct comparison.

Table 3-19—Available current in test circuits

Frame rating	Individual pole short-circuit test values			
	2-pole circuit breaker		3-pole circuit breaker	
	Amperes	Voltage	Amperes	Voltage
100 A max 250 V max	5000	L–L	4330	L–L
100 A max 251–600 V delta	10 000	L–L	8660	L–L
101–800 A delta voltage	10 000	L–L	8660	L–L
800 A max 480 Y/277 V or 600 Y/347 V	10 000	L–N	10 000	L–N
801–1200 A delta voltage	14 000	L–L	12 120	L–L
801–1200 A 480 Y/277 V or 600 Y/347 V	14 000	L–N	14 000	L–N
1201–2000 A delta voltage	14 000	L–L	14 000	L–L
2001–2500 A delta voltage	20 000	L–L	20 000	L–L
2501–3000 A delta voltage	25 000	L–L	25 000	L–L
3001–4000 A delta voltage	30 000	L–L	30 000	L–L
4001–5000 A delta voltage	40 000	L–L	40 000	L–L
5001–6000 A delta voltage	50 000	L–L	50 000	L–L

NOTES

Individual poles of multipole MCCBs are tested at short-circuit levels indicated in this table for all values of multipole interrupting ratings. These tests are in addition to multipole tests.

These are minimum test values for certification to UL 489-1991. They are not marked ratings and are printed here to aid the system designer who may need them for single-phase short-circuit analysis. Single-pole circuit breakers are tested at values equal to their interrupting rating.

L–L = line-to-line voltage applied.
L–N = line-to-neutral voltage applied.

Source: UL 489, October 9, 1991, Table 22.2.

Table 3-20—Short-circuit current tests

Test	Duty cycle	Type of test (no. of phases)	Rated maximum voltage	Current
1	O–15 s C-O	3	635	I_1
2	O–15 s C-O	3	508	I_2
3	O–15 s C-O	3	254	I_3
4	O–15 s C-O	1	635	0.87 I_1
5	O–15 s C-O	1	508	0.87 I_2
6	O–15 s C-O	1	254	0.87 I_3
7	O	3	635	I_1
8	O–15 s C-O	3	635	I_8
9	O	1	600	174 000
10	O+t+C-O	3	600	200 000
11	O	3	600	See 3.9.2.3 of ANSI C37.50-1989
12	O	3	600	See 3.9.2.4 of ANSI C37.50-1989

NOTES

1—O is the opening operation; C-O is close-open; t is the time necessary for the test procedures, including replacement of fuses and resetting of the open-fuse trip device; I_1 is the rated short-circuit current at rated maximum voltage of 635 V; I_2 is the rated short-circuit current at rated maximum voltage of 508 V; I_3 is the rated short-circuit current at rated maximum voltage of 254 V (see Table 1 of ANSI C37.16-1988); I_8 is the rated short-circuit current at rated maximum voltage of 635 V (see Table 2 of ANSI C37.16-1988).

2—Tests 1 and 2 are to be performed with opposite terminals energized. (e.g., if upper terminals are used for test 1, then lower terminals are used for test 2, and vice versa).

3—Test 2 is to be performed in sequence II given in Table 1 of ANSI C37.50-1989, using a circuit breaker equipped with the minimum-rated continuous-current electromechanical overcurrent trip device for the circuit-breaker frame size being tested.

4—Test 4, 5, and 6 may be performed on the same circuit-breaker, one test per pole.

5—For tests 9 and 10, the current is in rms symmetrical amperes.

6—For tests 11 and 12, see the appropriate sections of ANSI C37.50-1989.

7—At the option of the manufacturer, test 11 may be omitted if the total clearing time of the maximum fuse is equal to or less than the minimum total clearing time of the circuit-breaker element, at the short-circuit test current value. If the circuit breaker's time current characteristic data are for the maximum clearing time, subtract 0.016 s to obtain a value for the minimum total clearing time of the circuit-breaker element.

Source: ANSI C37.50-1989, Table 3.

Table 3-21—MCCB and ICCB short-circuit test summary

	Test no.	Tested in sequence no.	Duty cycle	No. of poles being tested	Max rated voltage	Actual test current rms symmetrical kA Frame rating (A)										
						100[a]	100[a]	225[a]	600[a]	800[a]	1200[a]	1600[a]	2000[a]	2500[a]	3000[a]	4000[a]
Standard tests	1	Z	O-CO	1	250	4.3	—	—	—		—	—	—	—	—	—
Standard tests	2	Z	O-CO	1	600	—	8.6	8.6	8.6	8.6	12.1	14	14	20	25	30
Standard tests	3	Z	O-CO	1	600	—	8.6	8.6	8.6	8.6	12.1	14	14	20	25	30
Standard tests	4	Z	O-CO	1	600	—	8.6	8.6	8.6	8.6	12.1	14	14	20	25	30
Standard tests	5	Z	O	3	250	5	—	—	—	—	—	—	—	—	—	—
Standard tests	6	Z	O	3	600	—	10	10	10	10	14	—	—	—	—	—
Standard tests	7	Z	O-CO	3	600	—	—	—	—	—	—	20	25	30	35	45
Standard tests	8	Y	O-CO	3	240	1.5	—	—	—	—	—	—	—	—	—	—
Standard tests	9	Y	O-CO	3	600	—	3	3	6	10	14	20	25	30	35	45

	Test no.[b]	Duty cycle	No. of poles	Trip rating	Actual test current
HIgh I/C tests	A[a]	O-CO	3	Maximum	Same as maximum interrupting capacity (I/C) rating
HIgh I/C tests	B[a]	O-CO	3	Maximum	I/C rating at maximum voltage rating
HIgh I/C tests	C[a]	O-CO	3	Maximum	I/C at maximum kVA rating
HIgh I/C tests	D[a]	O-CO	3	Maximum	Maximum I/C rating

[a]All tests at each rating in sequence Z must be sucessfully passed with single breaker.
[b]Each test may use a new breaker.

Source: [B13].

Table 3-22—LVPCB short-circuit test summary

Test no.	Tested in sequence no.	Duty cycle	No. of poles	Rated max voltage	Actual test current rms symmetrical kA						
					Frame rating (A)						
					225	600	800	1600	2000	3000	4000
1	I	O-CO	3	635	14	22	22	42	42	65	85
2	II	O-CO	3	508	22	30	30	50	50	65	85
3	II	O-CO	3	254	25	42	42	65	65	85	130
4	II	O-CO	1	635	12.2	19.1	19.1	36.5	36.5	56.6	74
5	II	O-CO	1	508	19.1	26.1	26.1	43.5	43.5	56.6	74
6	II	O-CO	1	254	21.8	36.5	36.5	56.6	56.6	74	113.1
7	III	O	3	635	14	22	22	42	42	65	85
8	IV	O-CO	3	635	14	22	22	42	42	65	85

NOTE—Any of the above tests may use a different breaker provided that the test sequence in progress is completed with the same breaker.

Source: [B13].

Table 3-23—UL and ANSI short-circuit test parameters

Parameter	UL 489		ANSI C37.50
Enclosure	Smallest individual		Min dimension test enclosure
Current	Per Table 5 of UL 489-1991		Per Table 5A of ANSI C37.50-1989
Voltage	100–105% rated		100-105% rated max (254 V, 508 V, or 635 V)
Power factor (ac)	10 000 or less 10 001–20 000 Over 20 000	0.45–0.50 0.25–0.30 0.15–0.20	0.15 max
Time constant (dc)	10 000 or less Over 10 000	3 ms 8 ms	Not covered
Frequency	48–62 Hz		48–72 Hz
Ambient	Not defined		Not defined
Ground fuse	30 A		30 A or 10 AWG (copper)

NOTES
1—Random closing employed.
2—Time interval between interrupting operations: 2 min–1 h max.
3—Cotton test during each test to 0.010 in diameter rod entry test must be passed.

Source: [B13].

Field testing by manual methods often produces results that are not in agreement with the manufacturer's tests and are not accurate indicators of circuit-breaker performance. It is difficult to justify a high-quality test setup anywhere except at a manufacturing facility, so differences between factory and field test results should be expected. Offset in the test current wave invalidates results.

When testing, it is always very important to make sure the trip unit installed is the one represented by the referenced specifications or time-current curves. Interchangeable trip units often look very similar but may be different. MCCBs, for example, may have thermal-magnetic trip units or electronic trip units. If they have electronic trip units, the electronic trip units may be peak sensing or rms sensing and they may or may not include ground-fault tripping provisions. Furthermore, newer electronic trip units feature built-in test provisions that are very easy to operate and can even be operated in a no-trip mode while the circuit breaker is under light load.

Alternative to using the built-in provisions, secondary or primary current injection methods can be employed using external test equipment. Current injection tests, of course, require the circuit breaker to be taken out of service. Thermal overload trip units utilizing bimetallic sensors respond to the rms value of the current flowing through them and their heaters. Electronic overload trip units may be either peak sensing or rms sensing depending upon their internal circuit design. That is, they can be designed to respond to the peak value of the current flowing through them or to the rms value of current flow. The engineer should always keep these factors in mind when evaluating alternative circuit-breaker and trip-unit choices.

The response of a bimetallic trip unit on circuit breakers will be different if the energy input to the trip unit as a whole is different. For example, if only one pole is carrying current, then only one bimetallic element is being heated and pressing on the trip bar and in general, a slightly longer trip time should be expected. This fact is indicated by the single-pole test characteristic printed above the long-time, time-current curve on typical circuit-breaker data sheets. The shorter time characteristic below it is indicative of performance, per standards, with equal current flow in all poles. See the time-current characteristic curves in Chapter 4.

Instantaneous electromagnetic pickup (tripping) in thermal-magnetic circuit breakers is in fact a function of peak current flow even though the abscissa of the time-current curve is labeled in rms amperes. This is so because the magnetic flux and the force the magnetic flux produces in the operating mechanism of the trip unit is a function of current alone, not power. In like manner, a peak sensing electronic trip unit can respond with a trip to a single current surge or spike that reaches the circuit trip level. Even electronic trip units using rms sensing algorithms can be triggered to produce an override instantaneous trip if a current surge or spike large enough to activate the override is experienced. The override trip is an independent instantaneous trip set near the circuit-breaker withstand level that overrides the electronic logic trip unit to cause the circuit breaker to open without delay at very large fault levels. Therefore, circuit-breaker application evaluations should take into account not only the salient features of the test specifications cited above, but also the requirements and characteristics of the circuit-breaker trip unit and the circuit breaker itself. Every aspect of circuit-breaker design, circuit operation, and system maintenance can affect overall operational performance.

3.34 Blow-open contact arms

Most MCCBs achieve high interrupting rating levels by their being designed to utilize fault current itself to drive tripping action. Within the limits of rating, it can be generalized that the larger the fault-current flow, the greater the driving force and the quicker the trip and interruption. Most MCCBs trip and interrupt fast enough to limit the peak and I^2t let-through of fault current. Some satisfy the requirements of current-limiting circuit breakers as defined in Chapter 2. LVPCBs, on the other hand, trip only after their trip units initiate a mechanism release. Then fault current can contribute driving force.

If an MMCB is claimed to be current limiting, UL 489-1991 requires that the peak current and I^2t be tabulated for the threshold of current limiting, the maximum interrupting level, and at least one point in between. A curve to present these data is usually drawn and published for user reference. Such a tabulation and curve are not required for circuit breakers not claimed to be current-limiting.

3.35 Circuit breaker useful life

It is prudent to replace any MCCB that has interrupted, at most, two faults at rated maximum current. This is so because the MCCB short-circuit proof test consists of an "O–t–CO" sequence, which means that in proof testing of the circuit-breaker design and in periodic follow-up testing thereafter, the circuit breaker is required to open a fault from an initially closed position (corresponding to the "O" operation), then after a period of time (t) to reset is allowed, to be closed into a maximum fault and trip open for a second time (corresponding to the "CO" operation). This demonstrates a circuit breaker's ability to perform at least two maximum level fault interruptions with the second at least a little worse than the first. No maintenance of the circuit breaker on test is permitted between interruptions.

The problem, of course, is that fault-current levels are not usually monitored. It is difficult and expensive to tell if a fault was a maximum fault and in general, low-voltage system faults tend to be less than maximum. Therefore, circuit-breaker inspections should be performed according to a plan developed to suit the application. NEMA AB 4-1996 should be referenced for MMCB and ICCB field inspection and maintenance.

LVPCBs go through similar short-circuit test cycles, but it is generally not said that LVPCBs need to be replaced after a given number of fault interruptions because they can and should be inspected for wear and damage and they should be refurbished or repaired as required after interrupting faults and before being restored to service. The fact that LVPCBs can even be maintained between tests emphasizes the maintainability feature of their design and further distinguishes them from MCCBs and ICCBs. See note (3) of Table 1 of ANSI C37.50-1981 for some specific detail on LVPCB testing. Maintenance is necessary if continued reliable service is to be expected.

3.36 Considerations on interrupting duty and maintenance

As discussed above, one problem associated with the implementation of good system operating and maintenance procedures is that it is generally difficult to determine if a fault that has occurred was a maximum level or bolted fault. Another factor is that without inspection, the actual condition of any circuit breaker cannot be known. Time and money must be spent to implement both procedures. Some new digital microprocessor-based trip units store fault-current magnitude data, both phase and ground, in memory when a trip occurs and that data can be read at the time of inspection. This helps the engineer determine the seriousness of a trip condition.

For the ultimate in reliability, the engineer should assume that the fault could have damaged any of the circuit elements, including the conductors, and a complete inspection of the circuit is required. MCCBs, ICCBs, and LVPCBs should be inspected in proportion to the required reliability of their service and as a minimum in observance of the recommendations given in standards and instruction leaflets. For a detailed discussion of MCCB inspection procedures, see NEMA AB 4-1996 in particular, and for a detailed discussion of circuit-breaker reliability in general, see IEEE Std 493-1990 [B8]. LVPCB, MCCB, and ICCB maintenance and inspection procedures can be found in the instruction leaflets and documentation furnished by circuit-breaker manufacturers. These documents should be read by system operating personnel upon receipt of the equipment and they should be kept on file for future reference. Maintenance personnel should incorporate pertinent practices and procedures into their own maintenance policies. The benefits of proof testing can be lost if inspection and maintenance policies are not implemented.

3.37 Integrally fused devices

Integrally fused LVPCBs and MCCBs with inverse time or instantaneous automatic tripping have interrupting capacities much greater than those of unfused circuit breakers of corresponding frame sizes, and are intended primarily for overcurrent and/or short-circuit protection of high-capacity electrical circuits. When applied on high short-circuit current capacity systems, the effects of the let-through characteristics of the fused circuit breakers on the connected equipment should be considered. The presence of the current-limiting fuse as part of the fused circuit breaker does not necessarily imply that the connected equipment can adequately withstand these effects.

It should be noted that fused circuit breakers do not generally have any current-limiting effect until the current associated with the fault exceeds the current-limiting threshold of the fuse. When fuses of relatively low continuous-current rating and relatively low peak let-through current rating are selected to give protection to downstream equipment, there is increased likelihood that they will open at currents much below the circuit-breaker element short-circuit current rating. If the full coordination study for the protection of connected equipment is made known to the manufacturer, then the best combination of direct-acting trip devices and fuses may be selected. Non-optimum combinations can lead to needless fuse opening. In no case should combinations of trip devices and fuses that are not approved by the manufacturer be installed.

Where fuses of different manufacture are being considered for the same system, the characteristics of all the fuses and circuit breakers in the system should be evaluated since both the melting time current characteristics and peak let-through currents of a given fuse rating may vary substantially between manufacturers. Only fuses that have been proof tested with the circuit breakers should be used.

3.38 Series-connected rating

Series connection of circuit breakers can be of economic advantage in an application only if full selectivity in coordination is not required. In strictest interpretation, domestic series combinations are also restricted to use in circuits where there is no motor loading on any of the branch circuit breakers. However, there are differences of opinion about this restriction in the U.S, and in other parts of the world. For example, IEEE Std C37.13 does not even consider the fault-current contribution of motors 50 hp and less and for a large part of the rest of the world, IEC 781 (1989) [B2] indicates that the contribution of asynchronous motors to the short-circuit current can be disregarded if the sum of the motor contribution is less than 1% of the initial symmetrical short-circuit current.

Sometimes it is erroneously thought that series combinations are at a disadvantage with regard to coordination as compared to fully rated systems. The fact is that even fully rated circuit breakers with instantaneous trips will not "coordinate" once the fault level exceeds both circuit breakers' instantaneous trip levels. An IEC viewpoint extends this concept somewhat by definition of the term "discrimination," which recognizes that energy is required to cause a circuit breaker to trip and even though the straight vertical lines and flat horizontal segments commonly used to describe the magnetic trip range of a circuit breaker are drawn, there is some range of overlap of these zones where tripping of both circuit breakers does not occur. The process of discrimination defines these areas so they can be added to the range of selectivity indicated by the time-current curves. The interested reader should refer to IEC 947-2 (1995) [B3] for more detail. Because there are enough opportunities to make series rating an advantage to users, series ratings and listings have been established by UL and are recognized in the NEC.

Series connection of MCCBs, where the branch or downstream circuit breaker has an interrupting rating less than the calculated fault duty at its point of connection in the power circuit, is permitted only when the series combination has been proven to be safe by actual interruption testing. Domestically, a series combination is recognized for series application by a third-party organization such as UL.

Series ratings should not be confused with the older domestic calculated cascade ratings discussed in 3.39. It should be noted that IEC uses the term *cascade* to describe its series rated and tested circuit-breaker combinations and the term *discrimination* to describe the ability of a load-side series-connected circuit breaker to actually coordinate with a line-side circuit breaker over some portion of their indicated mutual instantaneous trip range. Very fundamentally, series ratings are proven by test while the no longer valid cascade arrangements of the past were determined by calculation procedures that are no longer accepted as generally adequate.

NEC Section 110-22 acknowledges that manufacturers can establish series combination ratings and it requires that equipment enclosures "...shall be legibly marked in the field...Caution—Series Rated System" to indicate that the rules for series application were utilized to design this part of the power distribution system. This marking becomes part of the application. NEC Section 240-83 on Overcurrent Protection also acknowledges the use of series ratings and the requirements for marking. UL 489-1991 (in paragraph 41, Circuit Breakers Connected in Series) outlines the test connections and procedures required for proof of series combination ratings.

Series rating of two circuit breakers makes it possible to apply the two in series, as one device, with the interrupting rating being the series rating of the combination. In summary, it is not permissible to calculate series ratings because accurate and sufficiently uncomplicated methods for doing so have not been identified at this time.

3.39 Cascade arrangement

Previously in practice, there was a circuit-breaker arrangement know as a *cascade arrangement* in which circuit breakers were essentially applied in series. However, the adequacy of the cascade arrangement was determined by calculation, not by testing, and the calculation methods have since been determined to be generally inadequate. Since the cascading method does not include verification by testing, it is no longer a recommended procedure for applying circuit breakers. UL 489-1991 does not address the subject of cascade arrangements generally for MCCBs, while IEEE Std C37.13-1990 specifically states that it is no longer a recommended procedure for LVPCBs. If coordination considerations will permit the application of a series combination, then only tested and listed series combinations of circuit breakers can be applied and the markings of equipment discussed above should be employed. Otherwise, fully rated circuit breakers should be applied at all locations in the circuit with interrupting ratings equal to or greater than the evaluated prospective fault current at the point of application.

3.40 Short-time rating

Short-time ratings are not covered in MCCB standards. This is so because generally, MCCBs are designed to trip and interrupt high-level faults without intentional delay. However, newer electronic trip units usable with some MCCBs are able to utilize the capabilities of some of these circuit breakers to implement short-delay tripping. ICCBs generally do have short-time capability because their closing and tripping mechanisms are more like those of LVPCBs, not designed to blow open, and they generally have more current withstand capability. LVPCBs are designed to have “making current” capability and “short-time” capability and can withstand the short-time duty cycle test. They are designed to be tripped by a shunt trip device.

For an unfused LVPCB, the rated short-time current is the designated limit of available (prospective) current at which it shall be required to perform its short-time current duty cycle of two periods of 0.5 s current flow separated by a 15 s interval of zero current at rated maximum voltage under the prescribed test conditions. This current is expressed as the rms symmetrical value of current measured from the available current wave envelope at a time one-half cycle after short-circuit initiation.

Unfused LVPCBs shall be capable of performing the short-time current duty cycle with all degrees of current asymmetry produced by three-phase or single-phase circuits having a short-circuit power factor of 15% or greater (X/R ratio of 6.6 or less). Preferred short-time current ratings are shown in Table 3-9.

Fused circuit breakers do not have a rated short-time current; only the circuit-breaker element of the fused circuit-breaker assembly shall have such a rating and it shall be the same as described above.

3.41 Circuit-breaker evaluation for X/R ratio or short-circuit power factor

LVPCBs in general are evaluated for short-circuit interrupting capability on a first-half-cycle basis. As indicated previously, MCCBs can sometimes operate so quickly that they function in a current-limiting mode, which means they operate to limit short-circuit current before the first current peak is reached. Since the peak current is a function of the offset of the rms symmetrical current wave, which is in turn a function of the power factor or the X/R ratio of the circuit, the facts that (1) LVPCBs are tested with an X/R ratio of 6.6 and (2) MCCBs and ICCBs are tested with X/R ratios of 6.6 to 4.89, 3.8 to 3.18, and 1.98 to 1.75, depending upon interrupting rating, means they have to be evaluated differently. See Table 3-18 for a listing of the power factor ranges from which the MCCB X/R ratio ranges are derived.

If a circuit has an X/R ratio less than the value used for proof testing a given circuit breaker, then no change in that circuit breaker's interrupting rating is required and the circuit breaker can be evaluated by direct comparison of its short-circuit interrupting current rating with the calculated Ohm's Law symmetrical short-circuit fault-current calculation. Another way of understanding that statement is to understand that an increase in the short-circuit interrupting capability of a circuit breaker may never be claimed by virtue of a mathematical calculation alone.

On the other side of the X/R inequality, if the calculated value of the short-circuit X/R ratio is greater than the value used to test the circuit breaker, then the interrupting duty requirement of that application has to be increased by multiplying the calculated symmetrical short-circuit current magnitude by a multiplying factor (MF) greater than one, which is equal to the ratio of the offset peak of the calculated circuit divided by the offset peak of the test circuit. This means that the offset current for the calculated fault is greater than the offset current of the circuit-breaker test circuit and that the circuit breaker should therefore have the capability to interrupt MF times the calculated Ohm's Law value of symmetrical short-circuit current when applied in that circuit.

Looking at this from the point of view of de-rating the circuit breaker instead of up-rating the short-circuit current, it could be said that the circuit-breaker rated interrupting capacity should be de-rated by a factor equal to the reciprocal of MF (or 1/MF) because the peak fault current with this larger X/R condition is greater than the peak current of the circuit-breaker test circuit.

In summary, the calculated Ohm's Law symmetrical short-circuit current can be multiplied by an MF to indicate the true interrupting requirement of the circuit, or the short-circuit interrupting rating of the circuit breaker can be de-rated by multiplying its short-circuit interrupting rating by (1/MF) to indicate the circuit breaker's capacity to interrupt current on the new higher X/R basis. Both approaches are commonly used. See Tables 3-24 and 3-25 for multiplying factors.

Table 3-24—Selection of short-circuit current multiplying factor for LVPCBs

System short-circuit power factor (%)	System X/R ratio	Multiplying factor for calculated short-circuit current	
		Factors for unfused circuit breakers	Factors for fused circuit breakers
20	4.9	1.00	1.00
15	6.6	1.00	1.07
12	8.27	1.04	1.12
10	9.95	1.07	1.15
8.5	11.72	1.09	1.18
7	14.25	1.11	1.21
5	20.0	1.14	1.26

Source: IEEE Std C37.13-1990, Table 3.

3.42 Single-pole interrupting capability and power system design considerations

For the rated X/R condition, every three-pole circuit breaker intended for use on three-phase circuits is able to interrupt a bolted single-phase fault. Obviously, it should have this capability because single-phase faults not only cannot be prevented from occurring on three-phase systems, but they are probably the most likely to occur.

When interrupting a single-phase, line-to-line fault in a three-phase circuit, there are two circuit-breaker poles in series performing the interruption with line-to-line voltage impressed across the two poles in series. Theoretical maximum single-phase prospective fault current is therefore 87% of full three-phase bolted fault current. This interrupting duty is less severe than for a three-phase interruption test where the first pole to open can have a maximum of one and one-half times peak phase voltage impressed across that pole alone and the theoretical maximum three-phase prospective fault current is by definition 100%. Therefore, three-phase interruption tests also prove single-phase interrupting capability of three-pole circuit breakers.

However, if the three-phase power system is corner grounded, then a single-line-to-ground fault on the load side of the circuit breaker will result in single-phase fault current flowing through only one pole of the circuit breaker, but with line-to-line voltage impressed across that one pole. A review of the test specifications referenced previously will show that all LVPCBs are tested to prove single-pole interrupting capability at the 87% current level with maximum line-to-line voltage impressed across that one pole (see Table 3-20). But not all MCCBs and ICCBs receive the same 87% current, full line-to-line voltage single-pole test. All MCCBs and ICCBs are tested for single-pole performance at rated line-to-line voltage, but some are tested at lower than 87% of maximum fault-current level. As in all applications, it is necessary to calculate the interrupting requirement of the circuit and to apply a circuit breaker with the required interrupting rating. See Table 3-19 for the different individual levels.

Table 3-25—Selection of short-circuit current multiplying factor for MCCBs and ICCBs

Power factor (%)	X/R ratio	Interrupting rating (A)		
		10 000 or less	10 001 to 20 000	Over 20 000
		Multiplying factor		
4	24.98	1.62	1.37	1.23
5	19.97	1.59	1.35	1.22
6	16.64	1.57	1.33	1.20
7	14.25	1.55	1.31	1.18
8	12.46	1.53	1.29	1.16
9	11.07	1.51	1.28	1.15
10	9.95	1.49	1.26	1.13
11	9.04	1.47	1.24	1.12
12	8.27	1.45	1.23	1.10
13	7.63	1.43	1.21	1.09
14	7.07	1.41	1.20	1.08
15	6.59	1.39	1.18	1.06
16	6.17	1.38	1.17	1.05
17	5.80	1.36	1.15	1.04
18	5.49	1.35	1.14	1.02
19	5.17	1.33	1.13	1.01
20	4.90	1.31	1.11	1.00
21	4.86	1.31	1.11	1.00
22	4.43	1.28	1.09	1.00
23	4.23	1.27	1.08	1.00
24	4.05	1.26	1.06	1.00
25	3.87	1.24	1.05	1.00
26	3.71	1.23	1.04	1.00
27	3.57	1.22	1.03	1.00
28	3.43	1.20	1.02	1.00
29	3.30	1.19	1.01	1.00
30	3.18	1.18	1.00	1.00
35	2.68	1.13	1.00	1.00
40	2.29	1.08	1.00	1.00
45	1.98	1.04	1.00	1.00
50	1.98	1.04	1.00	1.00

Source: [B6].

Figure 3-1 shows the circuit connections for the tests of one-, two-, and three-pole MCCBs tabulated in the operations columns of Table 3-3. From these, the different connections used for straight voltage rated, slash voltage rated, and single-pole tests can be seen. It is therefore necessary to treat MCCB applications in corner-grounded systems differently from applications of LVPCBs. Generally, the circuit-breaker manufacturer should be consulted whenever corner-grounded system applications are involved.

For a wider perspective on this situation, IEC rated circuit breakers are not required to receive regular single-pole tests. The single-pole interrupting capacity aspect of performance is addressed in Appendix C of IEC 947-2 (1995) [B3] only for "... multi-pole circuit breakers intended for use on phase-earthed systems..." and then only at a prospective current "... equal to 25% of the ultimate rated short-circuit breaking capacity..."

Outside the situations of application in a corner-grounded system or double jeopardy on improperly operated, ungrounded, or high-resistance grounded systems, which require the occurrence of simultaneous bolted faults on the line side and the load side of a circuit breaker to get even the possibility of 87% current, single-pole interrupting capability has not been a major application factor worldwide over the last half century. Circuit-breaker sales literature notes cover the intentional corner-grounded system application contingency by noting that if the power system is corner grounded, then the purchaser should contact the factory for application assistance.

Power system design engineers should first determine the type of power system to be used with due consideration for its practical implementation. This means that if power systems are to be designed to be operated ungrounded or with high-resistance grounding, then it should be specified that they should be operated in accordance with the operating procedures set forth for such systems. For more detailed discussion on power system design considerations and their operation, the interested reader is referred to IEEE Std 141-1993 [B4].

3.43 References

This chapter shall be used in conjunction with the following publications:

ANSI C37.16-1988 (Reaff 1995), American National Standard for Switchgear—Low-Voltage Power Circuit Breakers and AC Power Circuit Protectors—Preferred Ratings, Related Requirements, and Application Recommendations.[4]

ANSI C37.50 (1981 and 1989 versions), American National Standard for Switchgear—Test Procedures for Low-Voltage AC Power Circuit Breakers Used in Enclosures.

IEEE Std C37.010-1979 (Reaff 1988), IEEE Application Guide for AC High-Voltage Circuit Breakers Rated on a Symmetrical Current Basis (ANSI).[5]

[4]ANSI publications are available from the Sales Department, American National Standards Institute, 11 West 42nd Street, 13th Floor, New York, NY 10036, USA.

IEEE Std C37.13 (1981 and 1990 versions), IEEE Standard for Low-Voltage AC Power Circuit Breakers Used in Enclosures (ANSI).[6]

IEEE Std C37.20.1-1993, IEEE Standard for Metal-Enclosed Low-Voltage Power Circuit Breaker Switchgear (ANSI).

NEMA AB 1-1993, Molded Case Circuit Breakers and Molded Case Switches (DoD).[7]

NEMA AB 4-1996, Guidelines for Inspection and Preventive Maintenance of Molded Case Circuit Breakers Used in Commercial and Industrial Applications.

NEMA CP 1-1988 (Reaff 1992), Shunt Capacitors.

NFPA 70-1993, National Electrical Code® (NEC®).[8]

UL 489-1991, Molded-Case Circuit Breakers and Circuit-Breaker Enclosures (DoD).[9]

3.44 Bibliography

Additional information may be found in the following sources:

[B1] Accredited Standards Committee C2-1997, National Electrical Safety Code® (NESC®) (ANSI).[10]

[B2] IEC 781 (1989), Application guide for calculation of short-circuit currents in low-voltage radial systems.

[B3] IEC 947-2 (1995), Low-voltage switchgear and controlgear—Part 2: Circuit-breakers.

[B4] IEEE Std 141-1993, IEEE Recommended Practice for Electric Power Distribution for Industrial Plants (*IEEE Red Book*) (ANSI).

[B5] IEEE Std 142-1991, IEEE Recommended Practice for Grounding of Industrial and Commercial Power Systems (*IEEE Green Book*) (ANSI).

---

[5]IEEE publications are available from the Institute of Electrical and Electronics Engineers, 445 Hoes Lane, P.O. Box 1331, Piscataway, NJ 08855-1331, USA.

[6]The 1990 version is available from the Institute of Electrical and Electronics Engineers, 445 Hoes Lane, P.O. Box 1331, Piscataway, NJ 08855-1331, USA; the 1981 version can be obtained from Global Engineering, 15 Inverness Way East, Englewood, CO 80112-5704, USA, tel. (303) 792-2181.

[7]NEMA publications are available from the National Electrical Manufacturers Association, 1300 N. 17th St., Ste. 1847, Rosslyn, VA 22209, USA.

[8]The NEC is available from Publications Sales, National Fire Protection Association, 1 Batterymarch Park, P.O. Box 9101, Quincy, MA 02269-9101, USA. It is also available from the Institute of Electrical and Electronics Engineers, 445 Hoes Lane, P.O. Box 1331, Piscataway, NJ 08855-1331, USA.

[9]UL 489-1991 includes replacement pages dated 27 Mar. 1990, 24 Oct. 1990, 9 Oct. 1991, and 30 Dec. 1991. UL standards are available from Global Engineering, 1990 M Street NW, Suite 400, Washington, DC, 20036, USA.

[10]The NESC is available from the Institute of Electrical and Electronics Engineers, 445 Hoes Lane, P.O. Box 1331, Piscataway, NJ 08855-1331, USA.

[B6] IEEE Std 242-1986 (Reaff 1991), IEEE Recommended Practice for Protection and Coordination of Industrial and Commercial Power Systems (*IEEE Buff Book*) (ANSI).

[B7] IEEE Std 399-1990, IEEE Recommended Practice for Industrial and Commercial Power Systems Analysis (*IEEE Brown Book*) (ANSI).

[B8] IEEE Std 493-1990, IEEE Recommended Practice for the Design of Reliable Industrial and Commercial Power Systems (*IEEE Gold Book*) (ANSI).

[B9] IEEE P551, Draft Recommended Practice for Calculating AC Short-Circuit Currents in Industrial and Commercial Power Systems (*Violet Book* Draft).

[B10] NFPA 70B-1994, Electrical Equipment Maintenance.

[B11] NFPA 70E-1995, Electrical Safety Requirements for Employee Workplaces.

[B12] Occupational Safety and Health Act (OSHA), U.S. Department of Labor, published in the *Federal Register.*[11]

[B13] Panel Discussion on Application of Molded-Case Circuit Breakers at *1991 IEEE IAS Annual Meeting,* Dearborn, MI. Figures and tables composed by R. O. D. Whitt of Westinghouse (*now* Cutler-Hammer).

[B14] UL 486A-1991, Wire Connectors and Soldering Lugs for Use with Copper Conductors (DoD).

[B15] UL 486B-1991, Wire Connectors for Use with Aluminum Conductors (DoD).

[B16] UL 486C-1991, Splicing Wire Connectors (DoD).

[B17] UL 486D-1993, Insulated Wire Connectors for Use with Underground Conductors.

[B18] UL 486E-1994, Equipment Wiring Terminals for Use with Aluminum and/or Copper Conductors.

[11]The Federal Register is available from the Superintendent of Documents, U.S. Government Printing Office, Washington, D.C. 20402.

Chapter 4
Specific applications

4.1 Scope

This chapter describes the systematic procedures for determining the type, rating, and protective characteristics of low-voltage circuit breakers applied for specific purposes. The three types of circuit breakers are as follows:

— Low-voltage power circuit breakers (LVPCBs)
— Molded-case circuit breakers (MCCBs)
— Insulated-case circuit breakers (ICCBs)

Specific circuit applications discussed in this chapter include the following:

a) Service entrance
b) Main circuit breakers
c) Tie circuit breakers
d) Feeder and branch circuit breakers:
    1) Cables
    2) Busway
    3) Switchgear, switchboards, panelboards, or motor control centers (MCCs)
    4) Motors
    5) Generators
    6) Capacitors
    7) Transformers

4.2 Selection considerations

Selection considerations should include the following:

a) Compliance with nationally recognized regulations and standards such as the National Electrical Code® (NEC®) (NFPA 70-1996),[1] and those from the Occupational Safety and Health Administration (OSHA), Underwriters Laboratories (UL), American National Standards Institute (ANSI), and National Electrical Manufacturers Association (NEMA), where applicable, along with any local codes or safety requirements. In addition, IEEE standards contain useful application information for various types of systems. These include IEEE Std 141-1993 for industrial plants, IEEE Std 241-1990 for commercial systems, and IEEE Std 242-1986 for protection and coordination of electrical systems.
b) Special or unusual requirements imposed by characteristics of the electrical power source.
c) Special or unusual requirements resulting from load characteristics.
d) Interconnected system performance objectives with respect to selective fault clearing.

[1]Information on references can be found in 4.12.

e) Unusual operating conditions.

f) Special requirements for personnel safety.

g) Type of equipment in which the circuit breaker is mounted (individual enclosure, panelboard, switchboard, MCC, metal-enclosed switchgear).

It is recognized that the type of facility (industrial plant, continuous process, commercial building, hospital, etc.), as well as economics, facility operating and maintenance philosophies and capabilities, and standardization programs may influence the selection process described, with particular effect on the type of circuit breaker being applied. Such aspects are, by necessity, excluded from consideration in this chapter.

Service conditions that differ from those described in Section 2 of ANSI C37.17-1979 and NEMA AB 3-1996 are beyond the scope of this chapter.

4.3 Selection approach for application requirements

This chapter covers the application of standard-purpose low-voltage circuit breakers in specific applications. Special-purpose circuit-breaker applications are covered in Chapter 6.

4.4 Selection approach for electrical ratings

4.4.1 System voltage

Circuit breakers are rated by voltage class and should be applied only to system voltages within their ratings. System voltage is a determining factor of the circuit-breaker interrupting rating. MCCBs have either straight voltage ratings or slash voltage ratings. Refer to 3.12. Circuit breakers with slash voltage ratings, such as 480Y/277 V or 120/240 V, may be applied only on solidly grounded neutral systems as shown in Figures 4-1 and 4-2. Circuit breakers with straight ratings, and all LVPCBs, can be applied on ungrounded as well as grounded systems.

4.4.2 System grounding

Most circuit breakers are rated for application on low-voltage systems that are solidly grounded, high-resistance grounded, or ungrounded. However, some MCCBs have reduced interrupting capability for a single pole, as indicated by a slash in the voltage rating. These MCCBs may only be applied on solidly grounded systems, as explained in 4.4.1.

When selecting the type of circuit breaker for use on low-voltage, solidly grounded systems of more than 150 V to ground, the use of integral ground-fault trip elements should be considered. Ground-fault tripping is recommended for some applications, and is required by the NEC for certain service entrance and feeder applications (refer to NEC Sections 239-95, 215-10, and 240-13).

Ground-fault trip elements may also be used on high-resistance grounded and ungrounded systems, but will not operate for the first ground fault. However, if a second ground fault occurs in a different phase in the system, they will provide backup protection, which is more sensitive than the phase elements.

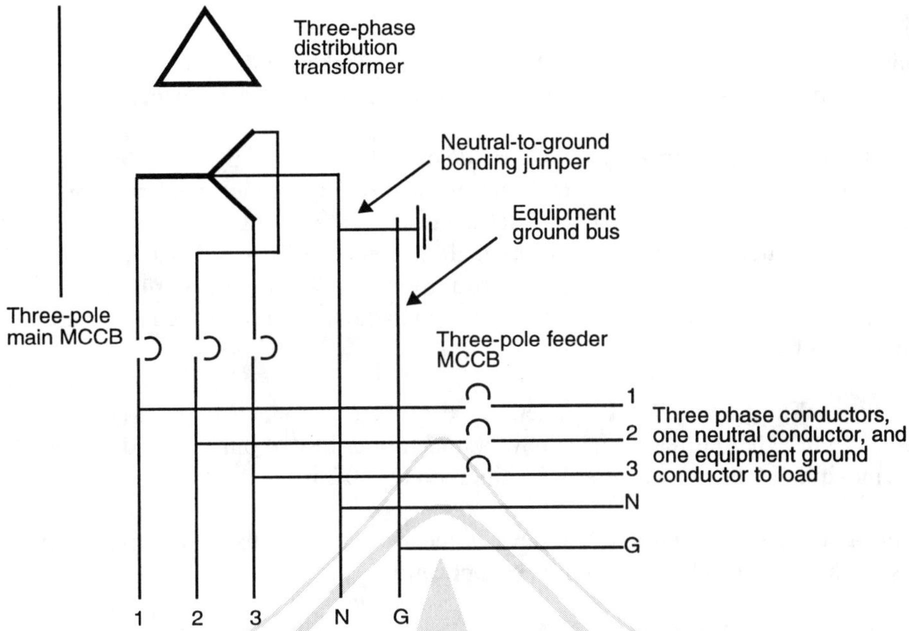

Figure 4-1—480Y/277 V power system

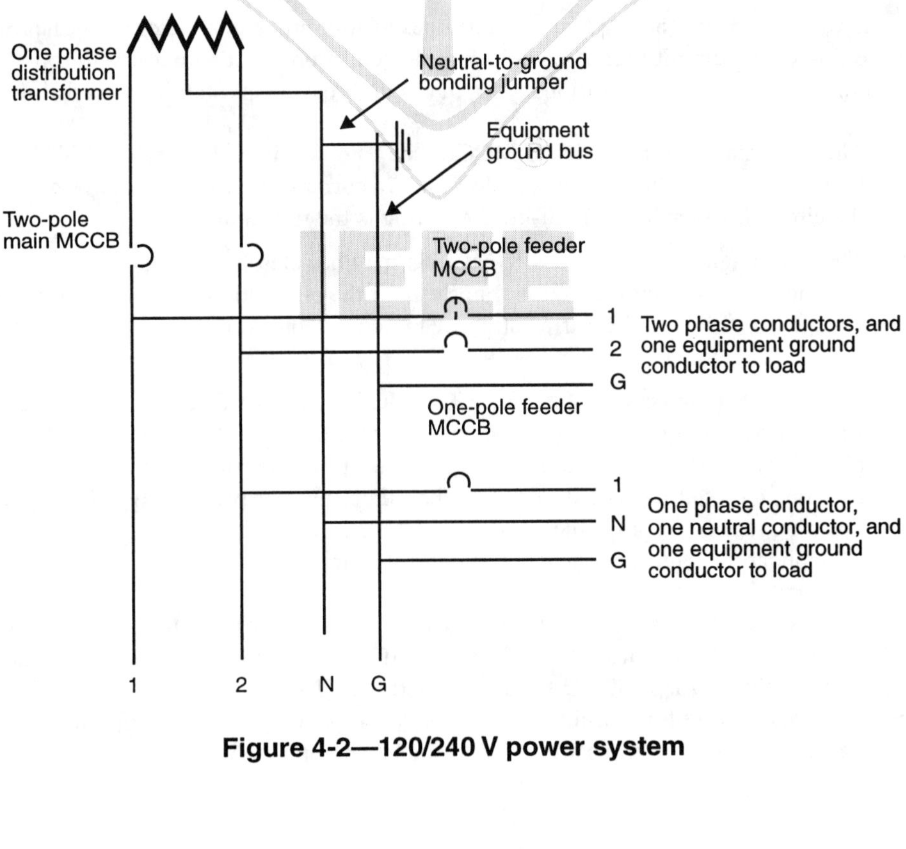

Figure 4-2—120/240 V power system

4.4.3 System frequency

Applications on systems other than 60 Hz should be checked with the manufacturer. Refer to 3.13. Systems rated 50 Hz may require special calibration of the trip device.

Circuit breakers with thermal-magnetic trip devices that are directly heated can generally be applied to power systems with a frequency up to 120 Hz without derating. Application of these circuit breakers above 120 Hz will result in increased eddy currents and iron losses, which cause greater heating within the thermal trip elements. To avoid this potential problem, the circuit breaker should either be calibrated for the specific frequency or be derated accordingly, or both. The amount of derating depends upon the frame size and current rating as well as the system frequency.

Some thermal-magnetic circuit breakers rated 600 A, and many thermal-magnetic circuit breakers with higher current ratings, have a transformer-heated bimetal, and are suitable for 60 Hz maximum. They require special calibration for 50 Hz.

Circuit breakers with electronic trip devices receive their signals from current transformer type sensors, and are calibrated for 60 Hz applications.

4.4.4 Continuous-load current

4.4.4.1 Continuous current rating or setting

The continuous-load current of a circuit determines the minimum conductor size. The trip rating or setting of the circuit breaker should be selected to protect the load and/or conductor. The trip rating or setting of the various types of circuit breakers is established as follows:

a) Thermal-magnetic trip units of MCCBs may be non-interchangeable. MCCBs containing these trip units are available in many current ratings up to the frame size of the circuit breaker. The circuit-breaker rating is the trip rating.

b) Thermal-magnetic trip devices of MCCBs may be interchangeable and are available in many current ratings up to the frame size of the circuit breaker. These units may be changed in the field. The trip rating of the circuit breaker depends on the trip unit installed.

c) Electronic trip units, available on MCCBs, ICCBs, and LVPCBs, use current sensors with various ratings equal to or less than the frame size of the circuit breaker. Rating plugs may be used to increase the range of settings. The trip unit provides an adjustable range of settings equal to or less than the sensor or plug rating. The sensor rating, plug rating (if applicable), and current setting selected from the adjustment range (if applicable) determine the trip setting of the circuit breaker.

Load current should not exceed the continuous-current rating or setting for "100% rated" circuit breakers, which includes all LVPCBs as well as ICCBs and MCCBs that are specifically rated and labeled for 100%. Other ICCBs and MCCBs may be applied at only 80% of the circuit-breaker rating for non-interchangeable trip type, or 80% of the trip unit rating for interchangeable trip type, or trip setting of adjustable trip type.

4.4.4.2 Ambient temperature and altitude

Derating of the circuit breaker's continuous current at higher ambient temperatures, humidity, or altitudes than rated conditions should be checked in 3.17 through 3.19.

4.4.4.3 Harmonics

Circuit breakers with trip devices that utilize rms sensing are most suited to applications where harmonics are known to be a problem. Peak sensing units react to the peak value of the distorted wave shape, which does not correspond to the effective heating value of the current.

4.4.5 Available short-circuit current

4.4.5.1 Interrupting rating

The interrupting rating of a circuit breaker is the highest current at rated voltage that it is intended to interrupt under standard test conditions. Refer to NEC Section 100-9, and 3.30 of this recommended practice.

The symmetrical interrupting rating of the circuit breaker shall exceed the calculated available short-circuit current at the point of application. The available short-circuit current includes contributions from all utility sources, plant generation, and connected motors. The interrupting rating of the circuit breaker is specific for the voltage at which it is applied. Refer to Tables 3-8 and 3-9 for standard interrupting ratings of the various types of low-voltage circuit breakers at different system voltages. Consult the manufacturers for ratings of specific circuit breakers, since some are available with interrupting ratings higher than the minimum rating required by ANSI C37.16-1988. A short-circuit study is required to determine the magnitude of three-phase and single-phase short-circuit current at various points in the system. The procedure for performing short-circuit studies is provided in IEEE Std 141-1993, IEEE Std 241-1990, and IEEE Std 242-1986. The calculated symmetrical short-circuit currents should be reviewed with respect to the expected system short-circuit X/R ratio or associated short-circuit power factor, because the interrupting rating of the circuit breaker is based on a specific maximum X/R ratio. This is described in 3.42.

To obtain selective coordination over the entire short-circuit current range, LVPCBs may be applied without instantaneous trip elements. When applied without the instantaneous trip element, the interrupting rating is the same as the short-time rating, as shown in Table 3-9. Consult the manufacturers for ratings of specific circuit breakers since some are available with interrupting ratings higher than the minimum rating required by the standards.

LVPCBs may be applied without integral trip units. In such cases, they are generally applied with separate overcurrent relays, and tripped using a shunt trip device. In these applications it is important to utilize the short-time rating rather than the interrupting rating, as shown in Table 3-9. The short-time current is defined as a maximum clearing time of 0.5 s at rated short circuit per ANSI C37.17-1979, Table 4.

The MCCB is available as a molded-case switch. This device has only an instantaneous element, which is designed to trip the breaker below its interrupting rating. As such, this attribute may cause coordination problems if an attempt is made to use it like a non-automatic circuit breaker.

When high interrupting ratings and/or current limiting capabilities are needed, current-limiting MCCBs or integrally fused MCCBs and LVPCBs may be used as a design option. Refer to Chapter 6.

4.4.5.2 Series-connected rating

The series-connected rating is a UL Recognized interrupting rating for a combination of line-side and load-side MCCBs. The load-side circuit-breaker interrupting rating may be less than the rating of the combination as shown in Figure 4-3. UL Recognized series combinations of fuses and MCCBs also exist (refer to [B1]).[2]

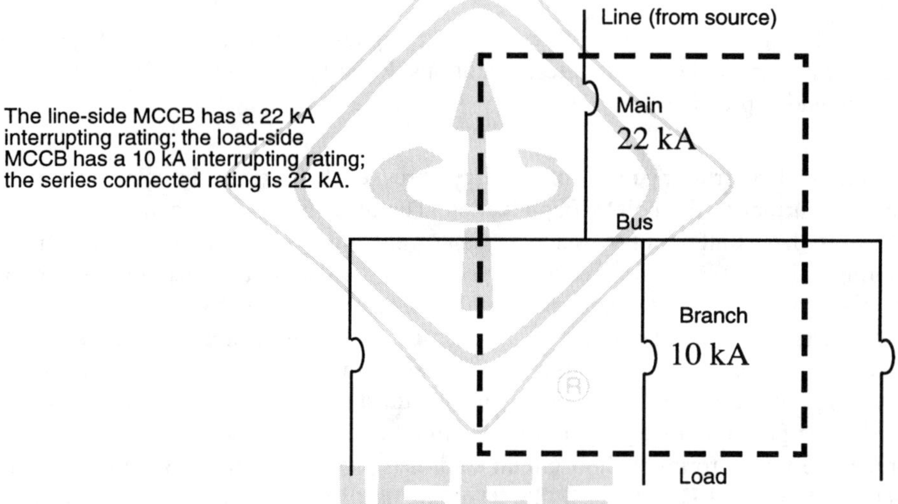

Figure 4-3—Series-connected rating

Equipment containing MCCBs, such as switchboards, panelboards, and residential service entrance equipment, must be tested and assigned a UL short-circuit rating when based on the rating of the series combination of circuit breakers used (tests may cover other main overcurrent protective devices).

Protective device series ratings are not limited to devices located in the same enclosure, such as panelboard main and branch circuit breakers. They can be located in different equipment, such as a residential metering distribution panelboard circuit breaker and a load-side residential load center, or a line-side switchboard and a load-side panelboard. Equipment will have rating labels that show short-circuit ratings when protected by series-connected rated line-side devices.

[2]The numbers in brackets correspond to those of the bibliography in 4.13.

The load-side circuit breaker of a series combination must be located in equipment that is listed and marked for use with series-connected ratings that include that circuit breaker.

Series-connected ratings for each manufacturer's equipment using series combinations of MCCBs are established by that manufacturer with testing witnessed by UL and the Canadian Standards Association (CSA). A series-combination should not use different manufacturers' circuit breakers, even though the manufacturers have similar designs, because no testing has been done to verify a series-connected rating.

The principal benefit of series ratings is the cost savings realized by using load-side circuit breakers whose interrupting rating is less than the available short-circuit current. However, there are disadvantages. The following should be considered when applying circuit breakers in a series combination:

a) One disadvantage of series combination is loss of selective coordination at high-fault currents. A fully-rated system might be arranged to avoid tripping the main circuit breaker for a feeder short circuit but the series combination requires both the feeder and main circuit breakers to trip when the available short-circuit current is above the instantaneous trip of the line-side circuit breaker. This is necessary to protect the load-side circuit breaker despite the disadvantage that opening the main circuit breaker interrupts power to all feeder loads that could have continued to operate in a selective system.

b) Series ratings require certain considerations in their applications that have to be handled by a power systems engineer. The line-side circuit breaker or other device opens to protect the underrated load-side circuit breaker when the short-circuit current exceeds the load-side circuit-breaker interrupting rating but is equal to or less than the line-side device rating. Both the line-side device and the load-side circuit breaker probably will trip for a large short circuit.

c) Series ratings cannot be applied if motors, or other equipment that contributes to a short-circuit current, are connected between the line-side MCCB or other device and the load-side MCCB.

d) In cases where increases in available short-circuit current necessitate a system upgrade, a second approach, shown in Figure 4-4, may be used for retrofitting existing older systems where a recognized series rating is not available. A line-side current-limiting circuit breaker or fuse, which limits peak current and let-through energy, may be added, only if the existing load-side breakers do not exhibit dynamic impedance within the first half cycle. The distribution of short-circuit energy is shifted away from the slower, load-side circuit breaker to the higher speed current-limiting device. The downstream circuit breaker is then subject to no more short-circuit energy than its rating. The manufacturer of the existing circuit breakers must be contacted to verify that they do not exhibit dynamic impedance.

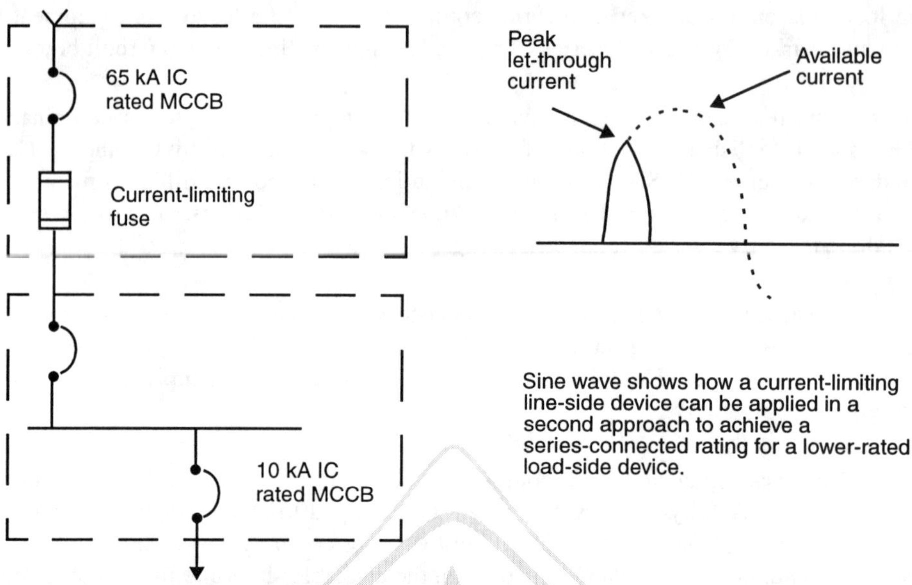

Figure 4-4—Series-connected rated devices

e) To accomplish selectivity, the circuit breakers shall have adjustable trip devices set to operate on the minimum level of short-circuit current. This permits them to be selective while distinguishing between short-circuit current and permissible load-current peaks. The circuit breakers should function in the minimum time possible and still be selective with other overcurrent protective devices in series. When these two requirements are met, the damage to equipment or the inconvenience caused by loss of power will be held to a minimum.

f) Series-connected ratings may be applicable when lowest first cost is the primary consideration and when selectivity, continuity of service, and lower maintenance costs are secondary considerations. When selectivity and reliability are more important than the first cost, the use of fully-rated equipment is recommended.

g) A portion of a specific manufacturer's UL listing of series-connected ratings for MCCBs is given in Table 4-1.

Table 4-1—Representative MCCB series-connected interrupting ratings

Main device			Branch breaker			Interrupting rating rms		
Type	A	Poles	Type	A	Poles	Symmetrical A	V ac	Phase
SKH	300–1200	2, 3	TK4V	400–1200	3	65 000	240	1, 3
SKH	300–1200	2, 3	TJJ	125–400	2, 3	65 000	240	1, 3
SKH	300–1200	2, 3	TJK	125–600	2, 3	65 000	240	1, 3
SKH	300–1200	2, 3	TJ4V	150–600	2, 3	65 000	240	1, 3
SKH	300–1200	2, 3	TFJ, TFK	70–225	2, 3	65 000	240	1, 3
SKH	300–1200	2, 3	TED	110–150	3	65 000	240	1, 3
SKH	300–1200	2, 3	TFJ, TFK	70–225	2, 3	25 000	480	1, 3
SKH	300–1200	2, 3	TJK	250–600	2, 3	35 000	480	1, 3
SKH	300–1200	2, 3	TJ4V	150–600	3	35 000	480	1, 3
SKH	300–1200	2, 3	TKM	300–1200	2, 3	35 000	480	1, 3
SKH	300–1200	2, 3	TK4V	400–1200	3	35 000	480	1, 3
SKH	300–1200	2, 3	TJJ	400	2, 3	35 000	480	1, 3
SKH	300–1200	2, 3	SFH	70–250	2, 3	35 000	480	1, 3
SKL	300–1200	2, 3	SFH	70–250	2, 3	100 000	240	1, 3
SKL	300–1200	2, 3	SFH	70–250	2, 3	65 000	480	1, 3
SKP	300–1200	2, 3	SFH, SFL	70–250	2, 3	200 000	240	1, 3
SKP	300–1200	2, 3	SFH, SFL	70–250	2, 3	100 000	480	1, 3
SKL	300–1200	2, 3	SKH	300–1200	2, 3	100 000	240	1, 3
SKL	300–1200	2, 3	SKH	300–1200	2, 3	65 000	480	1, 3
SKH	300–1200	3	SFH	70–250	3	25 000	600	1, 3
TPV	200–3000	3	SKH	300–1200	2, 3	100 000	240	1, 3
TB4	125–250	3	SED, SEH, SEL	15–150	2, 3	100 000	480	1, 3
TJJ	125–600	2, 3	SED	15–150	2, 3	25 000	480	1, 3
THFK	70–225	2, 3	SED	15–150	2, 3	25 000	480	1, 3
THED	110–125	2, 3	SED	15–150	2, 3	25 000	480	1, 3
THLC2	225	2, 3	SED, SEH, SEL	15–150	2, 3	200 000	480	1, 3
THLC2	225	2, 3	SED, SEH, SEL, SEP	15–150	2, 3	150 000	277	1
THLC2	225	2, 3	SED, SEH, SEL, SEP	15–150	2, 3	150 000	480	1, 3
THLC4	225–400	3	SED, SEH, SEL	15–150	2, 3	100 000	480	1, 3
THLC4	225–400	3	SED, SEH, SEL	15–150	2, 3	200 000	240	1, 3
THLC4	225–400	3	SED, SEH, SEL	15–150	2, 3	100 000	277	1
THLC4	225–400	3	SED, SEH, SEL, SEP	15–150	2, 3	150 000	480	1

NOTE—These ratings are specific to manufacturer, MCCB type, ampere, and voltage ratings.

Table 4-1—Representative MCCB series-connected interrupting ratings (*Continued*)

Main device			Branch breaker			Interrupting rating rms		
Type	A	Poles	Type	A	Poles	Symmetrical A	V ac	Phase
TEL	150	3	SED, SEH	15–150	2, 3	100 000	240	1, 3
TEL	150	3	SED, SEH	15–150	2, 3	65 000	480	1, 3
THLC4	225–400	3	SED, SEH, SEL	15–150	2, 3	200 000	120/240	1
TLB4	225–400	3	SED, SEH	15–150	2, 3	65 000	480	1, 3
TLB4	225–400	3	SED, SEH	15–150	2, 3	85 000	240	1, 3
TLB4	225–400	3	SED, SEH	15–150	2, 3	65 000	277	1
TLB4	225–400	3	SED, SEH	15–150	2, 3	85 000	120/240	1
NOTE—These ratings are specific to manufacturer, MCCB type, ampere, and voltage ratings.								

4.4.5.2.1 Example of a fully-rated versus a series-connected rated system

The following list discusses the difference between fully-rated and series-connected rated systems:

a) A fully-rated system has an available short-circuit current less than or equal to the short-circuit rating of the lowest rated component in the equipment. As shown in Figure 4-5, each protective device in a fully-rated system is rated to interrupt the available short-circuit current.

b) An alternative system uses a UL Listed panelboard containing UL Recognized series-rated combinations of circuit breakers whose individual short-circuit ratings may be below the available short-circuit current. For example, if it can be shown by test that a main circuit breaker with a 22 kA interrupting rating and a branch circuit breaker with a 10 kA interrupting rating will interrupt a 22 kA fault (see Figure 4-3), then the combination is series rated for 22 kA.

c) In the example series-connected rated system (see Figure 4-6), the combination of a 1600 A main circuit breaker with an interrupting capacity (IC) of 85 kA connected in series with either a 225 A circuit breaker with 25 kA IC or a 600 A circuit breaker with 42 kA IC has been tested and assigned a series-connected rating of 65 kA, which is less than the main circuit-breaker rating, but more than the rating of the feeder circuit breakers.

d) Regardless of whether the protective devices are fully rated or series rated, the bus bracing of the equipment must be equal to or exceed the available short-circuit current of the system as shown in Figures 4-5 and 4-6. An exception is when the line-side device is current limiting and tests have demonstrated a higher rating for the bus in series with the current-limiting device.

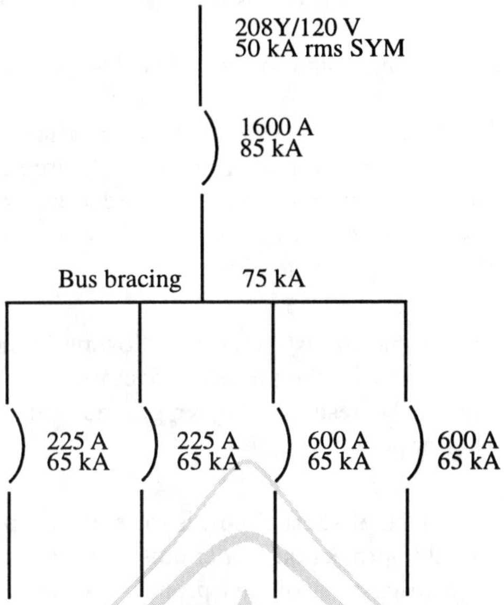

An example of a fully-rated system on a 208Y/120 V circuit with 50 kA symmetrical rms available; 1600 A main rated 85 kA IC connected to two 225 A and two 600 A breakers, all 65 kA IC.

Figure 4-5—Fully-rated system

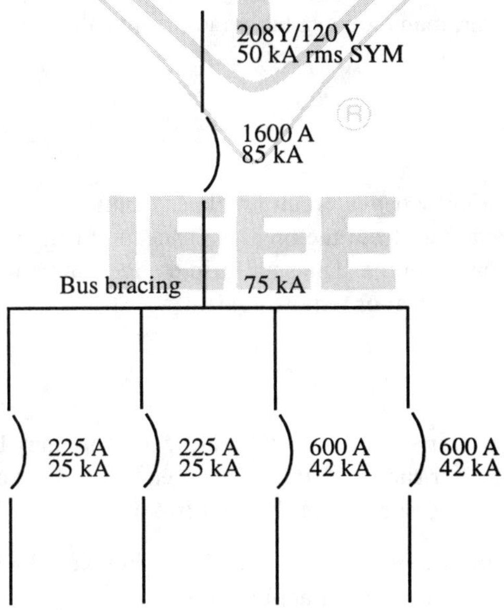

Combinations of 1600 A main breaker with 225 A or 600 A feeder breakers have been tested and assigned a series-connected rating of 65 kA IC. Equipment short-circuit rating is 65 kA.

Figure 4-6—Series-connected rated system

4.4.6 Arcing ground-fault protection for solidly grounded systems

In solidly grounded systems, the arcing line-to-ground short-circuit current is normally considerably lower than the value of a three-phase bolted fault. For example, in a 480Y/277 V system, arcing fault level can be as low as 38% of the three-phase value. However, arcing ground short-circuit currents can be much smaller due to high circuit impedance. Such short circuits, because of their destructive nature, must be removed as quickly as possible. Unfortunately, the magnitude of this current may be so low that low-voltage circuit-breaker long-time-delay characteristics allow it to persist too long.

Since it is the circuit impedance that limits the current flowing in an arcing short circuit, a ground conductor is often included with the phase conductors, to provide a lower reactance ground path for arcing current. The resulting higher ground-fault current is more readily detected and removed in a shorter period of time.

The best protection against this type of short circuit is to select a trip unit with a ground trip function; the second most suitable protection is a circuit breaker with a low-range instantaneous trip. (A more thorough discussion of this problem is given in Chapter 7 of IEEE Std 242-1986.) A ground-fault trip function may be required (refer to NEC Sections 230-95, 240-13, and 215-10).

4.5 Modifications and accessories for specific applications

Some modifications and accessories for low-voltage circuit breakers are available in kit form for field installation. However, many are available factory-installed and cannot be added later in the field.

4.5.1 Shunt trip device

A shunt trip is used to electrically trip a circuit breaker, manually or automatically, through a contact or switch located remotely from the breaker. The shunt trip circuit must be energized by some ac or dc control power source. The shunt trip device can be used for tripping from a separate protective relay, or for local or remote control.

When used for tripping from a protective relay

a) A reliable control power source should be utilized. This may be a station battery, or an uninterruptible power supply (UPS). If ac must be used, then it is necessary to also provide a capacitor trip device for each shunt trip device.

b) The short-time short-circuit rating of the circuit breaker shall exceed the available short-circuit current at the point of application.

When used for control purposes, the reliability of the control power source depends on the application. When used for remote control, the shunt trip device may be powered from the remote source.

4.5.2 Undervoltage release

An undervoltage release trips the circuit breaker whenever the voltage being monitored falls below a predetermined level. They are available with either time delay or instantaneous operation. The two types of undervoltage releases available are as follows:

a) *Electromechanical automatic reset used in combination with undervoltage release.* When the voltage falls below a predetermined level, a solenoid mechanism will initiate tripping of the circuit breaker. The circuit breaker cannot be closed until the voltage returns to approximately 85% of normal.

b) *Handle reset.* The handle reset undervoltage release spring is cocked or precharged through the circuit-breaker handle mechanism. The major advantage of this type is that the circuit-breaker mechanism cannot be latched when there is no power on the undervoltage release coil. This prevents circuit-breaker mechanism damage due to repeated attempts to close the circuit breaker with a de-energized undervoltage release coil.

NOTE—One of the design considerations for the use of both types of undervoltage release, as indicated above, is that they do not depend on control power to trip the circuit breaker.

The undervoltage release can be used to open the circuit breaker during a system undervoltage condition, in applications such as motor protection for cases where a magnetic contactor is not available to drop out, or where a sequenced restart is desired as opposed to full start-up of all devices and loads when power is restored.

4.5.3 Auxiliary switches

An auxiliary switch consists of "normally open" or "normally closed" contacts mounted in the circuit breaker that change state whenever the circuit breaker is opened or closed. To avoid confusion, the contacts are defined as "a" contacts that are *open* when the circuit breaker is open or tripped and "b" contacts that are *closed* when the circuit breaker is open or tripped.

Auxiliary switches may be used with an indicating device to show the position of the circuit breaker, and are used in the control circuit for interlocking purposes.

4.5.4 Mechanism operated cell (MOC) switch

Used with drawout circuit breakers, MOC switches are similar to auxiliary switches, except that they are mounted within a cubicle and are operated mechanically by the circuit-breaker mechanism. The "a" and "b" designations are the same as on auxiliary switches. They can be set to function in both the "Test and Connect" or the "Connect" only position. They are used when more auxiliary contacts are required than are available on the circuit breaker.

4.5.5 Truck operated cell (TOC) switches, cell switches, or position switches

Used with drawout circuit breakers, TOC switches change state when the circuit breaker is moved between the connected (operating) position and the test or disconnected position. A normally open contact is open when the breaker is *not* in the connected position. A normally closed contact is closed when the circuit breaker is *not* in the operating position. The TOC switch may be used with an indicating device to show the position of the circuit breaker, or in the control circuit to prevent operation of the circuit breaker in one of its positions.

4.5.6 Alarm switches

Alarm switches, sometimes referred to as bell alarm contacts, differ from auxiliary switches in that they function only when the circuit breaker trips automatically, not with the manual opening of the circuit breaker.

An alarm switch consists of a normally open and/or normally closed contact that changes state when the circuit breaker trips due to an overload, short-circuit, or ground fault. The contacts remain in this changed state until reset by a push button on the circuit breaker. Alarm switches are also available with an electrical reset mechanism that allows remote resetting.

An alarm switch may be used with an indicating device to show that the circuit breaker has tripped due to operation of the trip device, or it may be used in the control circuit to prevent closing of itself or another circuit breaker, until reset.

4.5.7 Motor operators on MCCBs

The motor operator, once it is activated by the remote push button or pilot device, will cause the circuit breaker to open or close. A motorized mechanism moves the MCCB handle from the tripped position to the closed position. This type of operation allows remote control of a circuit breaker. However, it is slow compared to the electrical close mechanisms on LVPCBs and ICCBs and may not be fast enough for applications such as synchronizing circuits.

4.5.8 Electrical close mechanism on LVPCBs and ICCBs

An electrical close mechanism consists of a stored energy closing mechanism, spring charging motor, and solenoid release. It usually includes anti-pump circuitry to prevent cycling. When the closing circuit is energized, the springs are charged. Once charged, operation of a remote push button, switch, or pilot device operates a solenoid that releases the springs, thus allowing the circuit breaker to close. After the circuit breaker is closed, the springs are recharged for the next trip-close-trip cycle of operations.

4.5.9 Mechanical interlocks

There are several methods of mechanically interlocking circuit breakers. These are walking beam, sliding bar, and key interlock. Each method results in the interlocking of two breakers so that only one may be closed (on) at the same time, yet both may be open (off) simultaneously. The type of interlock that may be used depends on the circuit breaker and the equipment in which it is mounted.

4.5.10 Moisture, fungus, and corrosion treatment

For an environment having a high moisture content or where fungus growth is prevalent, a special tropical treatment should be specified for the circuit breakers.

NOTE—Circuit breakers should not be exposed to corrosive environments. If there is no alternative, specially treated circuit breakers that are resistant to corrosive environments should be specified.

4.5.11 Terminal shields

Terminal shields protect personnel from accidental contact with energized current-carrying parts.

4.5.12 Handle locks

Handle locks are available to prevent accidental or deliberate manual operation of the circuit breaker. The lock does not prevent opening of the circuit breaker by its trip device.

4.5.13 Handle ties

Handle ties are used to connect two or more circuit-breaker handles together to enable manual operation of all poles simultaneously. The handle tie does not prevent opening of the circuit breaker by its trip device.

4.5.14 Shutters

Shutters are used to provide isolation from the primary contacts when the circuit breaker is withdrawn. They are not circuit-breaker accessories, but are used in equipment with circuit breakers. Shutters are applied only with draw-out circuit breakers.

4.6 Normal versus abnormal conditions

Normal environmental and operating conditions are as follows:

— Ambient temperature between 0 °C and 40 °C
— Altitude does not exceed 6600 ft (2000 m)
— Seismic zone 0
— Frequency of 60 Hz

Abnormal environmental and operating conditions that should be considered are as follows:

— Operation at ambient temperatures below 0 °C or above 40 °C
— Operation at altitudes above 6600 ft (2000 m)
— Exposure to corrosive materials
— Exposure to explosive fumes or dust

- Exposure to dust or moisture
- Seismic zone 1, 2, 3, or 4
- Abnormal vibrations
- Unusual operating duties
- Harmonics
- Repetitive duty cycle, which results in several operations in a short period of time on a regular basis
- Capacitor bank switching
- Frequent switching
- Circuits with high X/R ratios
- Single-pole interruption with three-pole circuit breakers
- Frequencies other than 60 Hz
- Occurance of frequent and/or severe faults

4.7 Considerations for applying MCCBs, ICCBs, and LVPCBs

Certain significant design, construction, or testing differences between low-voltage circuit breaker types may determine the choice of circuit-breaker type to be selected. ICCBs are considered to be a type of MCCB, and are tested in accordance with MCCB standards unless specifically stated otherwise by the manufacturer. However, ICCBs do have some of the features of LVPCBs. Refer to Table 4-2 for a comparison of circuit-breaker features. Table 4-2 can be helpful in the selection process.

Table 4-2—Comparison of features

LVPCB	ICCB	MCCB
Selective trip over full range of fault currents up to interrupting rating.	Selective trip over partial range of fault currents within interrupting rating.	Selective trip over a smaller range of fault currents within interrupting rating.
Type of operators: mechanically operated, two-step stored energy, and electrical two-step stored energy.	Types of operators: mechanically operated, two-step stored energy, and electrical two-step stored energy.	Type of operators: mechanically operated over-center toggle or motor operator.
Available in draw-out construction permitting racking to a distinct "test position" and removal for maintenance.	Available in draw-out construction permitting racking to a distinct "test position" and removal for maintenance.	Some are available in plug-in design allowing removal for inspection and maintenance. Large frame sizes may be available in draw-out construction.
Operation counter is available.	Operation counter is available.	Operation counter is available.
Interrupting duty at 480 V ac: 22–100 kA without fuses and up to 200 kA with integral fuses.	Interrupting duty at 480 V ac: 22 –100 kA.	Interrupting duty at 480 V ac: 22–65 kA without fuses and up to 200 kA with integral fuses or for current-limiting type.
Current limiting available only with fuses.	Current limiting not available.	Current limiting available with and without fuses.
Usually most costly.	Usually mid-range cost, but depends on the enclosure selected.	Usually least costly.
Small number of frame sizes available.	Small number of frame sizes available.	Large number of frame sizes available.
Extensive maintenance possible on all frame sizes.	Limited maintenance possible on larger frame sizes.	Limited maintenance possible on larger frame sizes.
Used in enclosures, switchgear, and switchboards.	Used in enclosures, switchgear, and switchboards.	Used in enclosures, panelboards, and switchboards.
Not available in series ratings.	Not available in series ratings.	Available in series ratings.
100% continuous-current rated in its enclosure.	80% continuous-current rated, unless specifically stated to be rated 100% in an enclosure.	80% continuous-current rated, unless specifically stated to be rated 100% in an enclosure.
IEEE Std C37.13-1990	UL 489-1991	UL 489-1991

4.8 Service requirements and protection

NEC Sections 230-70 through 230-95 contain the many requirements for services of 600 V or less including the sizing, location, and overcurrent protection of conductors, disconnecting means, permissible number of disconnects, rating of disconnects, grounding of conductors, and ground-fault protection requirements of service equipment.

4.9 Main circuit breakers

The main circuit breaker, as shown in Figure 4-7, is used for switching, servicing, and protecting the main bus of an assembly of low-voltage equipment, such as a line-up of switchgear, or a switchboard, panelboard, or MCC. It is often an integral part of the assembly, but can be separately located from the distribution assembly, if desired. When part of a service entrance, it is the service disconnecting means, as defined in the NEC. When the circuit breaker is located in the secondary of a stepdown transformer, it serves as the transformer secondary main circuit breaker, and should be located as close to the transformer terminals as possible. A main circuit breaker is not always mandatory, but the advantages it provides should be considered.

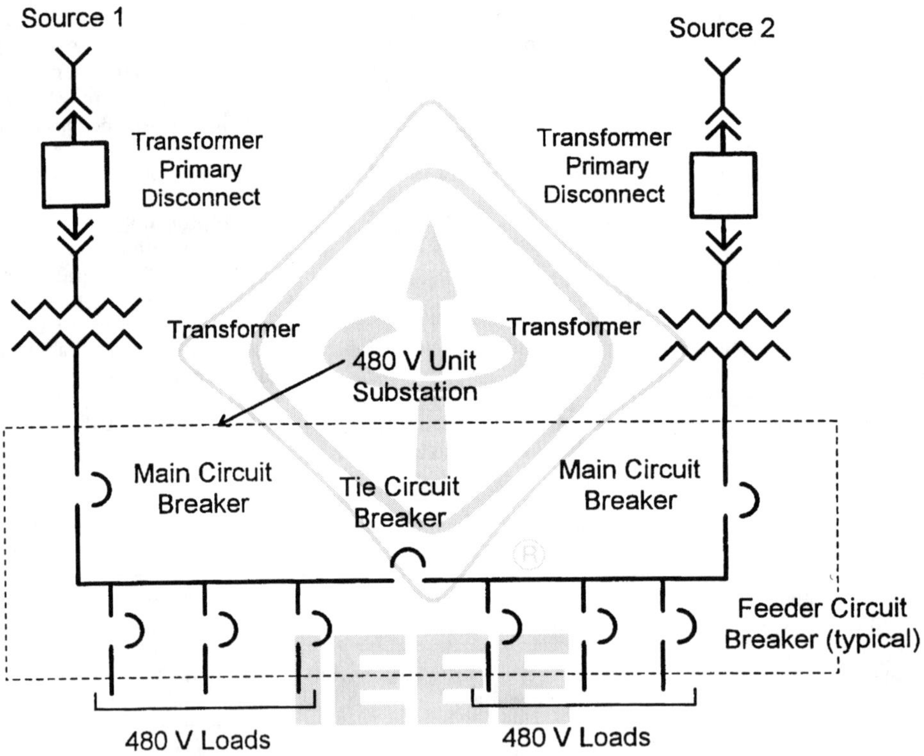

Figure 4-7—Typical double-ended unit substation

4.9.1 Disconnecting means

Opening the main circuit breaker isolates the load from the power source. It is used to de-energize the system, and is very useful when it is necessary to quickly turn off the power, such as when a fire occurs in the facility. For this reason it is mandatory in a service entrance application, if the service has more than six feeders or branches. It is useful during maintenance of the equipment to safely lockout/tagout everything downstream, when inspecting and maintaining the main bus and connections. The main breaker is also useful for re-energizing the system in an orderly fashion.

4.9.2 Protection device

4.9.2.1 Overload protection

The main circuit breaker provides overload protection for the main bus of the distribution equipment, as well as for the incoming power conductors to the circuit breaker. If applied at the transformer secondary, the main circuit breaker provides overload protection for that transformer. The main normally provides better overload protection than the transformer primary protective device. The primary device has a higher current rating or setting to prevent the device from tripping on transformer inrush current during transformer energization.

4.9.2.2 Short-circuit protection

The main circuit breaker provides short-circuit (fault) protection for the conductors (cable and/or bus) between the main circuit breaker and the branch or feeder circuit breakers. For example, it provides short-circuit protection for the main bus in the assembly, as well as for the tap-offs to the branch or feeder devices.

4.9.2.3 Ground-fault protection

This optional protection is very desirable for the main circuit breaker on solidly grounded systems of more than 150 V to ground because of the possibility of low-magnitude arcing ground faults that can occur in the main bus bars which are normally not insulated. Ground-fault protection may be required by the NEC (refer to NEC Sections 230-95 and 240-13).

4.9.2.4 General application considerations

General application considerations for the main circuit breaker include the following:

a) The preferred trip functions for selective trip are long-time and short-time (and ground fault, if required). For coordination purposes, instantaneous should be provided only if necessitated by the circuit-breaker interrupting rating.

b) May require key interlocking with a high-voltage switch on the transformer primary.

c) May require key or electrical interlocking with a tie circuit breaker.

4.10 Tie circuit breakers

The tie circuit breaker, as indicated in Figure 4-7, is used for switching, servicing, and protecting the main bus of an assembly of low-voltage equipment, such as a line-up of switchgear, or a switchboard, panelboard, or MCC. It is often an integral part of the assembly, but can be separately located from the distribution assembly, if desired. It is also used for sectionalizing or isolating a section of bus and to allow for maintenance of the main circuit breaker or transformer. A tie circuit breaker is never mandatory, but the advantages it provides should be considered. The functions performed by the tie circuit breaker include those described in 4.10.1 and 4.10.2.

4.10.1 Disconnecting means

Opening the tie circuit breaker along with one of the main circuit breakers isolates the included bus section from the power source. The tie circuit breaker is used to de-energize a portion of a system, and is very useful when it is necessary to quickly turn off the power. Opening one main circuit breaker and closing the tie circuit breaker enables the system to remain energized while the main circuit breaker or transformer is being maintained. It is useful during maintenance of the equipment to safely lockout/tagout everything downstream, when inspecting and maintaining the main bus and connections. The tie circuit breaker is also useful for re-energizing the system in an orderly fashion.

4.10.2 Protection device

4.10.2.1 Overload protection

The tie circuit breaker provides overload protection for a portion of the main bus of the distribution equipment, as well as for the upstream power conductors.

4.10.2.2 Short-circuit protection

The tie circuit breaker provides short-circuit (fault) protection for the conductors between the tie circuit breaker and the branch or feeder circuit breakers on that portion of the bus. For example, it provides short-circuit protection for the main bus in the assembly, as well as for the tap-offs to the branch or feeder devices.

4.10.2.3 Ground-fault protection

This optional protection is very desirable for the tie circuit breaker on solidly grounded systems of more than 150 V to ground because of the possibility of low-magnitude arcing ground faults that can occur in the main bus bars, which are normally not insulated. This may limit the number of feeders that are de-energized in a fault condition (refer to 4.4.6). A ground-fault trip function may be selected to provide ground-fault coordination with the main and feeder circuit breakers (refer to NEC Section 240-12).

4.10.2.4 General application considerations

General application considerations for the circuit breaker include the following:

a) The preferred trip functions for selective trip are long-time and short-time (and ground fault, if required or if desired, for coordination). For coordination purposes, instantaneous trip functions should be provided only if necessitated by the circuit-breaker interrupting rating.

b) May require key or electrical interlocking with main breakers to prevent paralleling or synchronism check equipment to monitor paralleling.

c) On four-wire multisource systems, ground-fault protection is complex and requires careful consideration.

4.11 Feeder protection

Feeder circuit breakers, like the main and tie circuit breakers, function as a disconnecting means for various types of loads as described in the paragraphs below. The feeder circuit breaker contains a protective device with the trip functions [overload, short circuit, and ground fault (if required)]. Feeder circuit protection is also covered in great detail in Chapter 8 of IEEE Std 242-1986.

4.11.1 Feeder circuit protection

4.11.1.1 Overload protection of cables

Overload protection is covered in NEC Section 240-3 under the provision requiring all conductors to be protected in accordance with their ampacity. In general, the ampacity of cables is determined from the tables contained in NEC Section 310-15, which concern the installation of conductors. The tables in NEC Section 310-15 also offer rules for derating cables. In general, no specific NEC rules are presented to coordinate insulation heating characteristics with the overcurrent devices concerning temperature versus time.

Overload protection cannot be applied until the ampacity of a cable is determined.

The normal ampacities of cable under the jurisdiction of the NEC are tabulated in its current (1996) issue or amendments thereto. The ampacity of cables under general operating conditions that may not come under the jurisdiction of the NEC are published by the Insulated Cable Engineers Association (ICEA). See 8.5 of IEEE Std 242-1986.

Derating may be required; refer to NEC Section 310-15.

Short-time overloading may be permissible in emergencies; details are examined in Chapter 8 of IEEE Std 242-1986.

4.11.1.2 Short-circuit protection

Cable conductors must be protected from overheating due to excessive short-circuit current. The task of providing cable protection during a short-circuit condition involves obtaining the following information:

a) Maximum available short-circuit currents

b) Maximum operating temperature of the insulation

c) Maximum conductor temperature that will not damage the insulation

d) Cable conductor size and material affecting I^2R value and the capability to contain heat

e) Longest time that the short circuit will exist and the short-circuit current will flow (see Table 4-3)

f) Cable short-circuit current damage curve (see Figures 4-8 through 4-10)

This subject is discussed in detail in 8.4 of IEEE Std 242-1986.

Table 4-3—Estimated clearing times of low-voltage circuit breakers

LVPCBs		
	Frame size	
	225–600 A	**1600–4000 A**
Instantaneous, cycles	2–3	3
Short time, cycles	10–30	10–30
Long time, seconds	over 100 s	—
Ground fault, cycles	10–30	10–30

MCCBs		
	Frame size	
	100 A	**225–4000 A**
Instantaneous, cycles	1.1	1.5

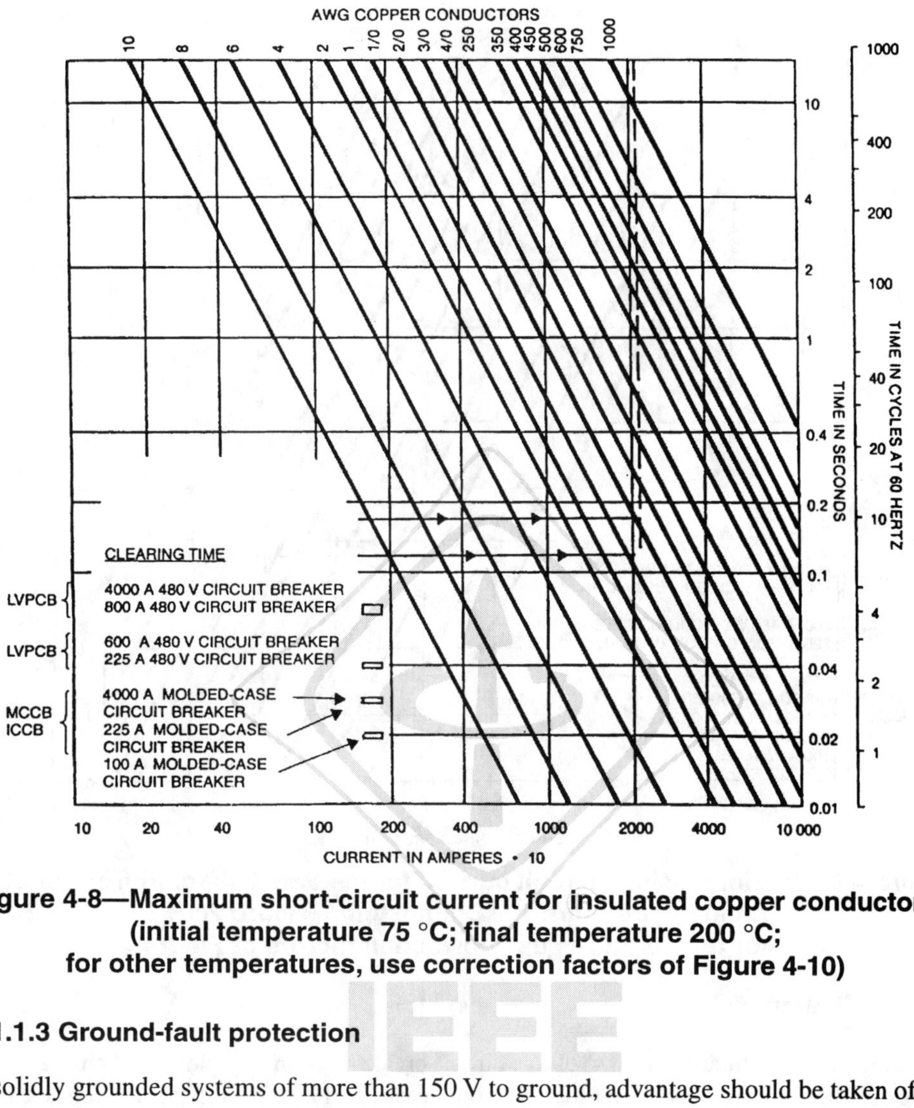

Figure 4-8—Maximum short-circuit current for insulated copper conductors (initial temperature 75 °C; final temperature 200 °C; for other temperatures, use correction factors of Figure 4-10)

4.11.1.3 Ground-fault protection

On solidly grounded systems of more than 150 V to ground, advantage should be taken of the enhanced protection offered by integral ground-fault trip devices available on most LVPCBs, ICCBs, and MCCBs. Phase-to-ground short circuits are more likely to occur than other types of short circuits. An integral ground trip provides sensitive protection with its low pickup current, which is usually a percentage of the circuit-breaker long-time-delay pickup or current sensor rating (refer to 4.4.6). A ground-fault trip function is selected to provide ground-fault coordination with the main and tie circuit breakers and coordination with the load protective devices. Where several steps of coordination are required, it may be necessary to ignore the residual method for determining ground-fault current, and instead choose an external source that measures the true ground-fault current, such as a current transformer that encircles all current-carrying conductors or a current transformer that measures current in the power transformer neutral-to-ground conductor (refer to NEC Section 240-12). A separate ground-fault protective device may be required for feeders rated 1000 A or higher (refer to NEC Section 215-10).

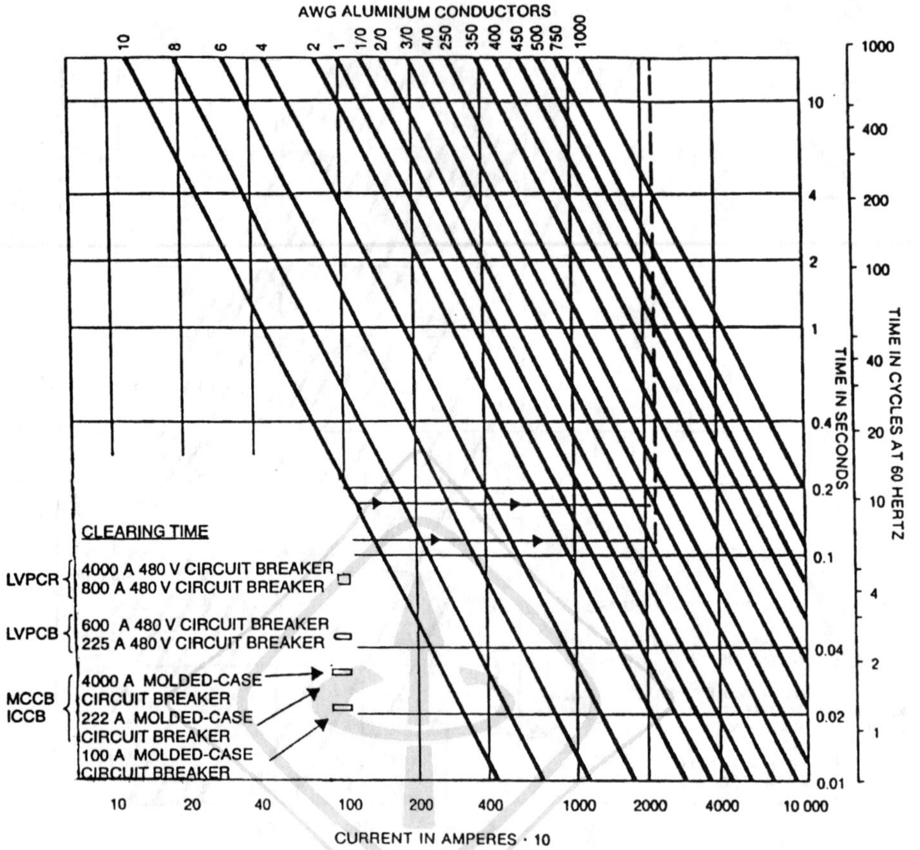

Figure 4-9—Maximum short-circuit current for insulated aluminum conductors (initial temperature 75 °C; final temperature 200 °C; for other temperatures use correction factors of Figure 4-10)

4.11.1.4 Example

Selecting and coordinating a low-voltage circuit breaker is done by plotting the time-current curves of the protected cable and the low-voltage circuit breaker on the same log-log graph paper (refer to Chapter 5).

The time-current curve of the circuit breaker should always be below and to the left of the maximum short-circuit current-damage curve (see Figures 4-11 and 4-12) of the protected cable. Figures 4-11 and 4-12 illustrate that a 600 V, Rated, No. 4/0 AWG copper insulated cable may be protected by an instantaneous tripping function of an LVPCB (see Figure 4-11) or MCCB (see Figure 4-12).

NEC Section 220-10 requires that, for a continuous load, a circuit breaker must be rated at least 125% of the continuous load. For a noncontinuous plus continuous load, a circuit breaker must be rated not less than the noncontinuous load plus 125% of the continuous load. An exception is if the circuit breaker is listed for operation at 100% of its rating, the ampere rating may be selected at not less than the sum of the noncontinuous load plus the continuous

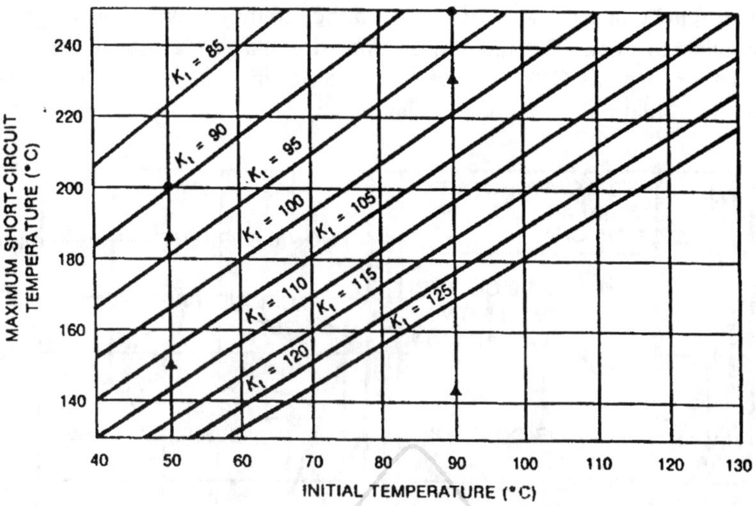

Figure 4-10—Correction factors (K_t) for initial and maximum short-circuit temperatures

load. All LVPCBs and only some MCCBs are 100% rated and may be applied at 100% of the continuous load. The load of Figure 4-12 is assumed to be noncontinuous and a 225 A MCCB is applicable. The thermal or long-time characteristic plotted on Figure 4-12 is less than the 230 A allowable cable ampacity. Refer to NEC Table 310-16 for 75 °C cable insulation.

Electronic trip devices provide better protection than thermal-magnetic trips. The MCCB provides short-circuit protection because the MCCB curve is below and to the left of the cable's maximum short-circuit current (damage) curve. For safe cable protection, the cable should be selected and protected in the same manner as described in the preceding paragraph.

4.11.2 Protection of busway

A busway may be considered a single replacement for several smaller cables. To contribute to the high reliability expected, it should be provided with the best possible protection, to minimize the number and duration of outages.

The two basic busway types are feeder and plug-in busway. Both are available in variations providing higher or lower impedance and with differing losses, enclosures, ventilation, or insulation features. See 8.8 of IEEE Std 242-1986.

4.11.2.1 Typical busway protective device

Like any other circuit, busways are subject to overloads and bolted short circuits. In addition, busways are subject to arcing ground faults on solidly grounded systems having line-to-ground voltages greater than 150 V. While no single element incorporates all the necessary characteristics, it is possible to assemble several elements into a single device. A particularly effective device is the fused, or current-limiting circuit breaker equipped with ground-trip protection. In such a device, the circuit-breaker elements provide operation in the overload and low short-circuit current range while the coordinated current-limiting capability functions during high short-circuit currents. The ground-trip sensor detects those arcing short-circuit currents that go to ground and, even though they may be low magnitude, signals the circuit breaker to open. The integral ground trips that are available on most LVPCBs, ICCBs, and larger size MCCBs provide fast and sensitive protection; the ground current pickup is usually a fraction of long-time delay pickup or sensor or plug rating.

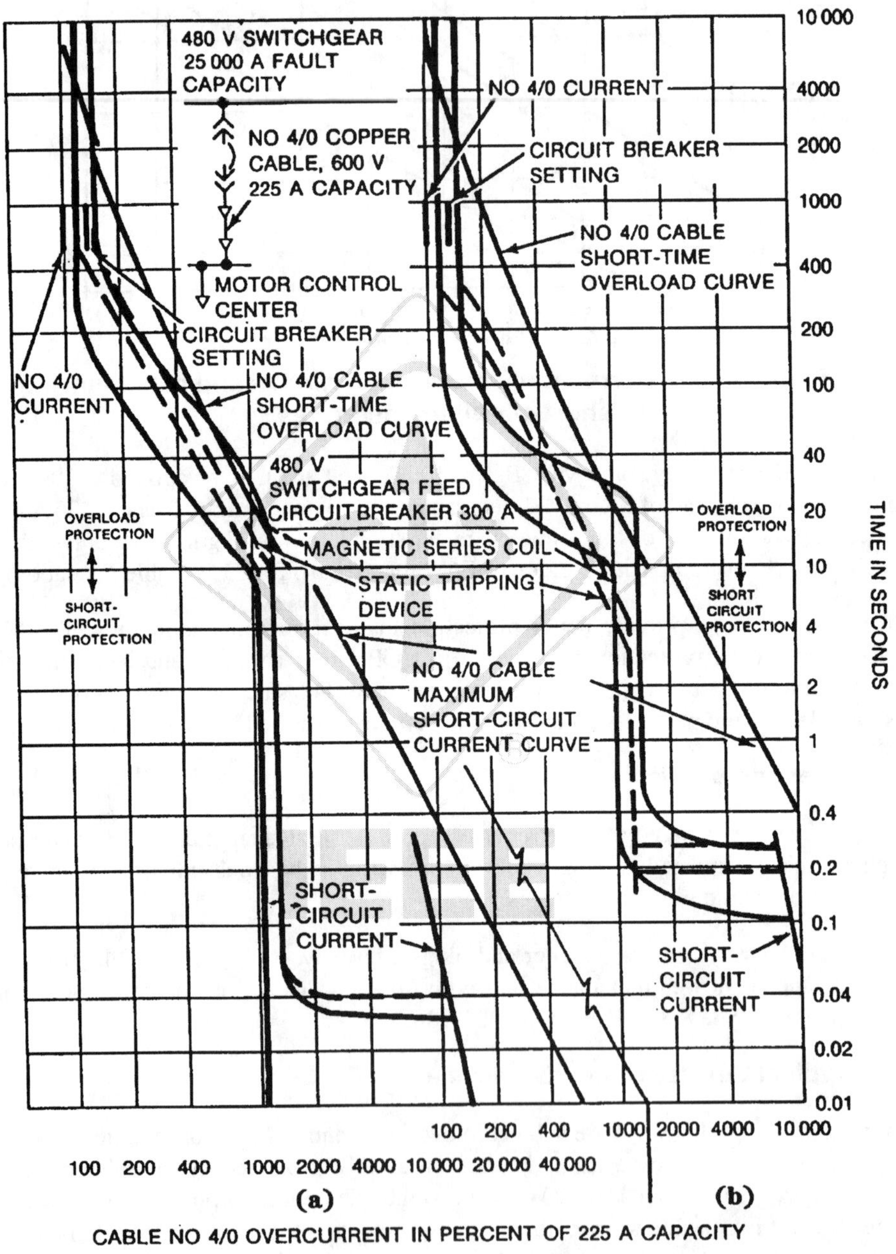

Figure 4-11—Short-circuit and overload protection of 600 V cables
(a) Long-time and instantaneous equipped LVPCBs
(b) Long-time and short-time equipped LVPCBs

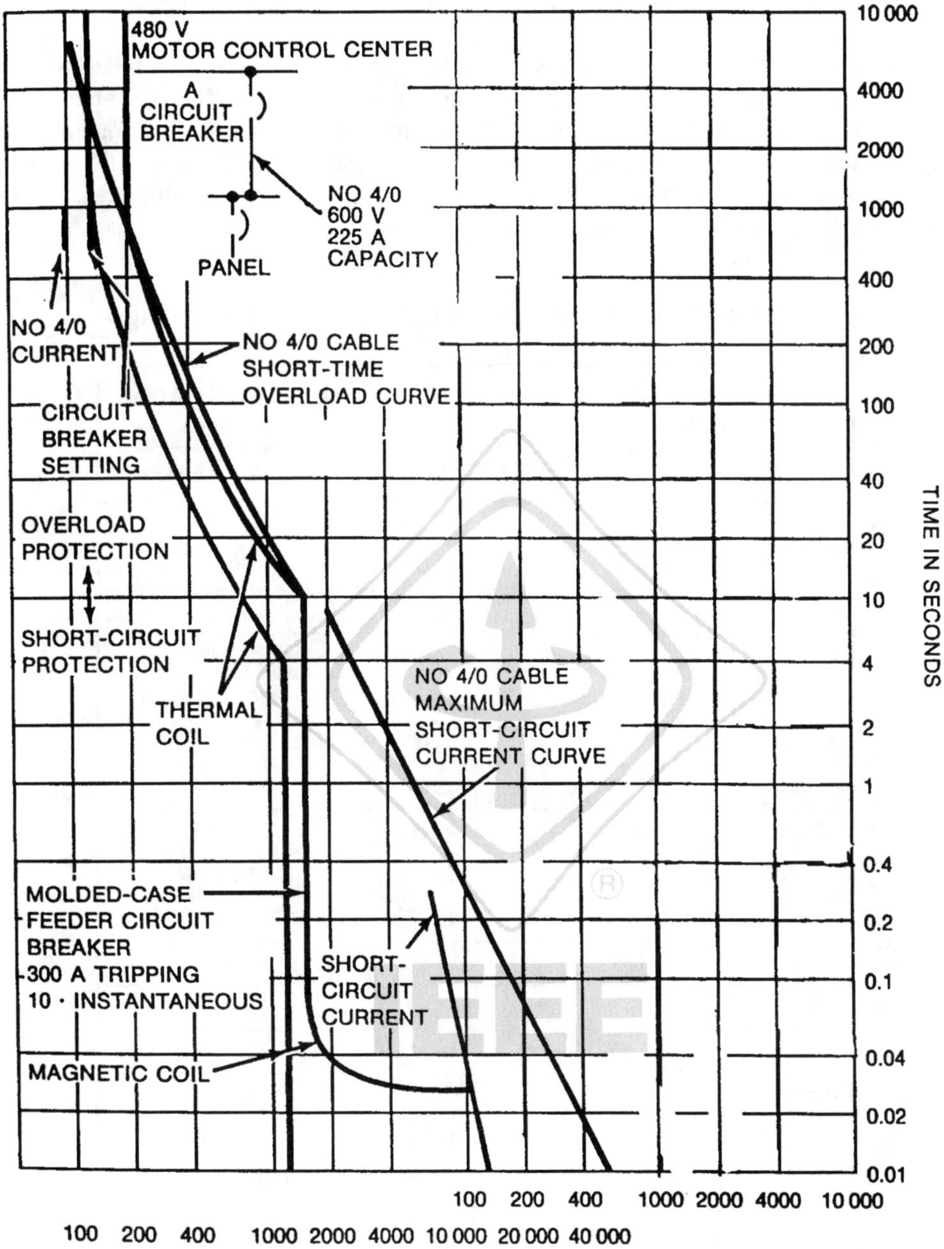

Figure 4-12 Short-circuit and overload protection of 600 V cables—Thermal-magnetic MCCBs

4.11.2.2 Busway thermal and mechanical high-current capabilities

Busway has a thermal overload and short-circuit withstand capability similar to cable. Table 4-4 lists busway minimum short-circuit current ratings. The time-temperature conserved-heat formulas for short-time high currents previously specified for cable also apply to busway. Thermal withstand lines on a log-log plot similar to those of Figures 4-8 and 4-9 can be calculated knowing an initial temperature and the permissible transient total temperature for the busway insulation.

Table 4-4—Busway minimum short-circuit current ratings

Continuous-current rating of busway (A)		Minimum short-circuit current ratings (A)	
Plug-in	Feeder	Symmetrical	Asymmetrical
100	—	10 000	10 000
225	—	14 000	15 000
400	—	22 000	25 000
600	—	22 000	25 000
—	600	42 000	50 000
800	—	22 000	25 000
—	800	42 000	50 000
1000	—	42 000	50 000
—	1000	75 000	85 000
1350	—	42 000	50 000
—	1350	75 000	85 000
1600	—	65 000	75 000
—	1600	100 000	110 000
2000	—	65 000	75 000
—	2000	100 000	110 000
2500	—	65 000	75 000
—	2500	150 000	165 000
3000	—	85 000	100 000
—	3000	150 000	165 000
—	4000	200 000	225 000
—	5000	200 000	225 000

The busway short-circuit rating, however, defines a mechanical limit that is lower than the thermal capability. This mechanical limit, therefore, applies for high currents below but near the short-circuit rating. Permissible flow times for these high currents, longer than the three cycles at 60 Hz (0.05 s) required at rating, are obtained from a constant I^2t mechanical limit characteristic.

For currents below one-half of the short-circuit current rating, where stresses reduce to one-quarter of those at rating, the mechanical limit becomes less important and the thermal capability determines protection requirements. The thermal capability at high current is constant I^2t and this joins the busway continuous-current rating through a smooth transition.

In addition, the damage due to arcing ground faults should be considered for solidly grounded systems with line-to-ground voltages greater than 150 V (see 8.8.1.5 of IEEE Std 242-1986).

4.11.2.3 Examples

The following two examples illustrate busway phase overcurrent protection by low-voltage circuit breakers whose time-current characteristics are below the busway thermal-mechanical capability characteristics. See Figures 4-13 and 4-14.

In the first example (Figure 4-13), an 800 A plug-in busway has a 22 kA short-circuit rating. It is assumed that the busway is applied where the bolted short-circuit duty is equal to the 22 kA busway rating, indicated on Figure 4-13 by B enclosed in a triangle. The LVPCB protecting the busway must clear this maximum bolted short-circuit current in the three cycle busway permissible time and must therefore be instantaneously tripped. For high currents, this circuit breaker cannot be selectively coordinated with busway plug fuses or circuit breakers that are instantaneously tripped. The busway protecting circuit-breaker instantaneous pickup is set as high as possible, without crossing the busway limit curve, and coordination with busway plug devices is achieved for short-circuit currents lower than this pickup. The short-time characteristic provides relatively fast tripping for currents in the arcing fault region, at A enclosed in a triangle and below. The curves of Figure 4-13 illustrate coordination with a 400 A busway plug MCCB having an instantaneous pickup of 10×, equalling 4000 A. If the largest busway plug device has a lower instantaneous pickup, the LVPCB short-time pickup should be correspondingly reduced to provide better short-time protection for arcing faults.

The long-time pickup of the protecting LVPCB should be no more than the next higher rating, in conformance with the requirements of NEC Sections 364-10 and 364-11. For solidly grounded systems of more than 150 V to ground, the electronic trip should also be equipped with a ground-trip function to provide fast, sensitive protection for arcing ground short circuits. The ground-trip pickup can be set at a fraction of the busway rated current.

An ICCB or large size MCCB with electronic trips can provide essentially the same busway protective characteristic as the LVPCB of this example.

The obvious disadvantage of the first example is the loss of selective coordination between the busway plug MCCB and the busway protection LVPCB. A severe short circuit on the plug circuit could trip both the plug and the LVPCB and interrupt other healthy circuits plugged into the busway. This disadvantage is removed by using busway with a higher short-circuit rating, as shown in the second example (Figure 4-14). As shown in Figure 4-14, the 800 A plug-in busway has a 65 kA short-circuit rating.

The LVPCB protecting the busway does not have an instantaneous trip and therefore coordinates selectively with busway plug fuses and circuit breakers that are instantaneously tripped. In other respects, the comments of the first example also apply to the second. If an ICCB were used instead of the LVPCB, the ICCB would still have a short-time and instantaneous unit as in the first example. Thus there would be a loss of selective coordination between the busway plug MCCB and the busway protection ICCB.

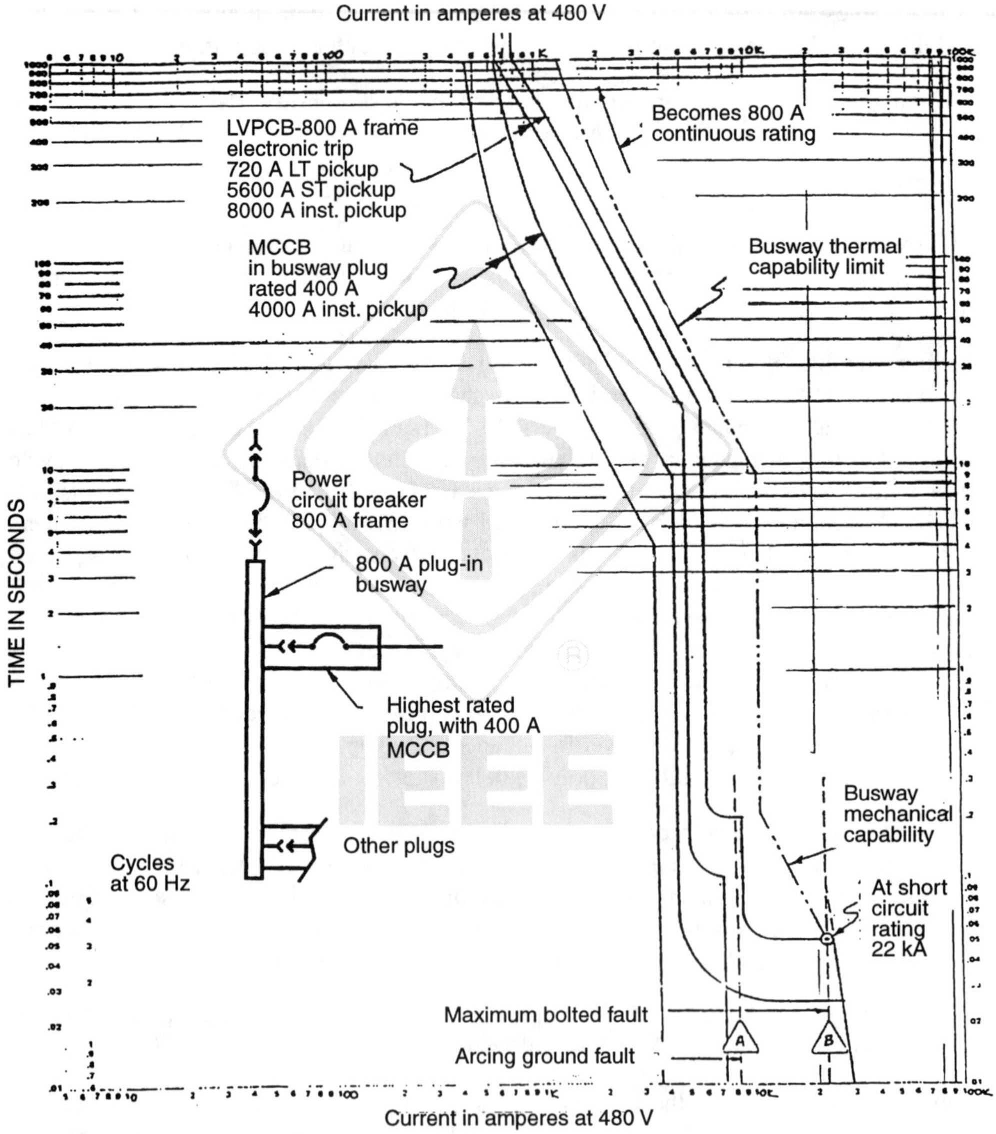

Figure 4-13 Short-circuit and overload protection of busway 22 kA short-circuit rating

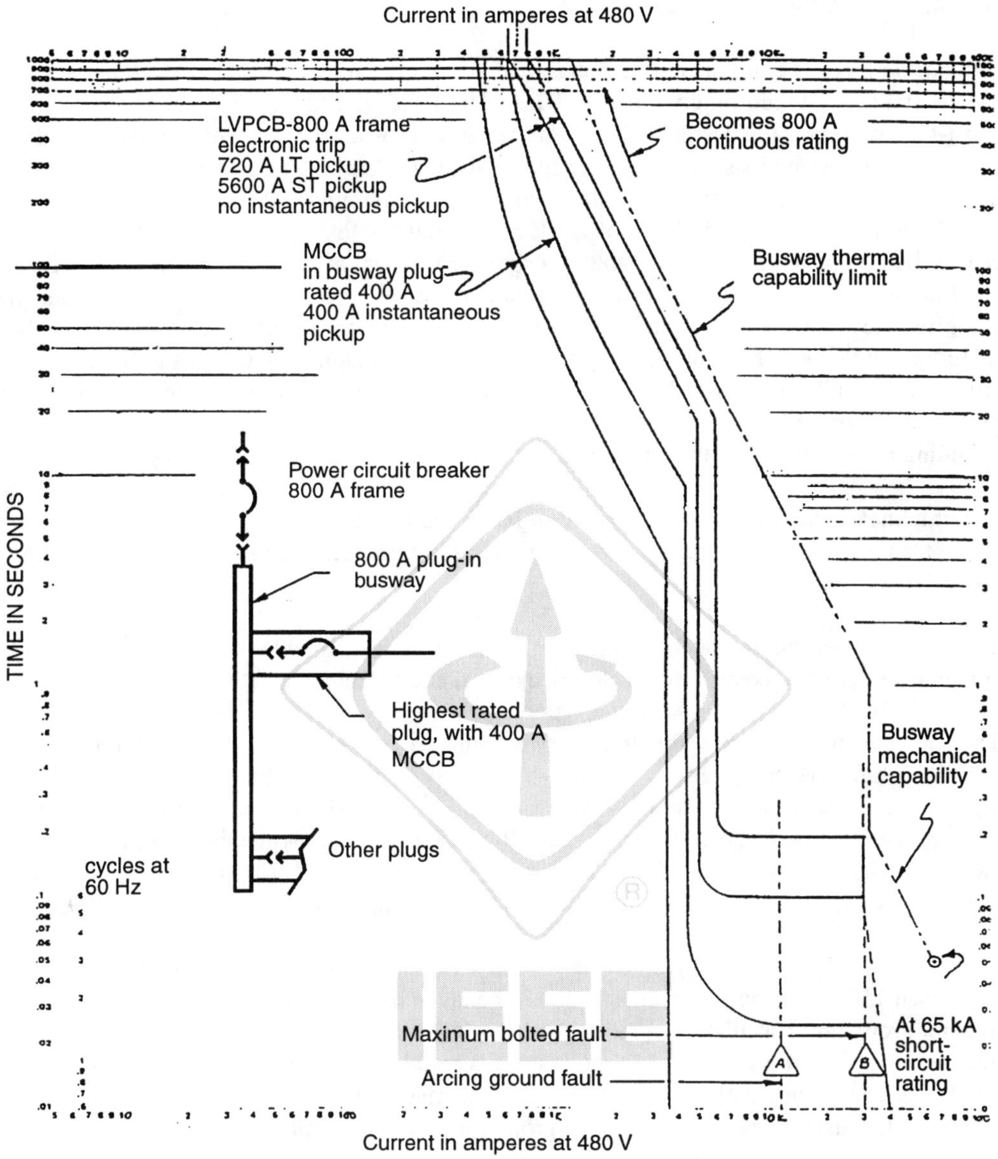

Figure 4-14 —Short-circuit and overload protection of busway 65 kA short-circuit rating

In summary, the short-circuit rating of the busway should be selected as high as possible in order to achieve selectivity.

The examples in Figures 4-13 and 4-14 show how an alternative selection of equipment short-circuit capabilities, along with trip unit characteristics/functions, can change a nonselective system to a selective system.

4.11.2.4 Protection of switchgear bus

Switchgear bus is a component of a prefabricated switchgear assembly manufactured according to standards. Accordingly, as stated in 4.4.3 of IEEE Std C37.20.1-1993, "The rated short-time current is the designated limit of available (prospective) current at which it shall be required to withstand its short-time duty cycle (two periods of 0.5 s current flow, separated by a 15 s interval of zero current) at rated maximum voltage under the prescribed test conditions." Similarly, as stated in 4.4.4 of IEEE Std C37.20.1-1993, "The rated short-circuit current is the designated limit of available (prospective) current at rated maximum voltage that it shall be required to withstand for a period no less than 4 cycles on a 60 Hz basis under the prescribed test conditions." Both are first-cycle rms symmetrical current ratings. Assemblies including switchgear bus shall be capable of withstanding these duties with all degrees of current asymmetry produced by three-phase or single-phase circuits having a short-circuit power factor of 15% or greater (X/R ratio 6.6 or less). The ratings are equal to the corresponding ratings of the smallest frame size circuit breaker used in the assembly.

Separate metal-enclosed buses described in IEEE Std C37.20.1-1993 have withstand ratings that match or exceed the interrupting ratings of circuit breakers having comparable continuous-current ratings.

These ratings determine that switchgear bus will not be damaged by currents flowing while connected circuit breakers are interrupting external short circuits.

The source or tie circuit breakers that carry currents to the switchgear bus may be set to provide long-time thermal overload and short-circuit protection for the bus, but the outgoing feeder circuit breakers act to provide protection against short-circuit currents that flow through the bus to faults external to the bus. Because of the short-time thermal ratings of the switchgear bus, the main, tie, and feeder protective devices on LVPCBs may have a short-time delay of up to 0.5 s at maximum short-circuit current, per IEEE Std C37.20.1-1993.

For sensitive tripping of arcing phase-to-ground short circuits on solidly grounded systems of more than 150 V to ground, advantage should be taken of the enhanced protection offered by integral ground trips available on most LVPCBs, ICCBs, and on larger size MCCBs. Switchgear arcing phase-to-ground short-circuit currents could be small with respect to the main circuit breaker long-time pickup and, without the sensitive ground-trip protection, might remain undetected, causing severe damage. The ground trips act fast and have low pickup currents, usually a fraction of the long-time pickup setting.

4.11.2.5 Protection of switchboard bus

Switchboard bus is a component of a prefabricated switchboard assembly manufactured according to standards. According to UL 891-1994 and NEMA PB 2-1995, switchboards are tested for three cycles at a power factor of 50%, 30%, or 20% for short-circuit ratings of 10 000 A, 10 001–20 000 A, or 20 001–200 000 A rms symmetrical current, respectively. As a result of this, LVPCBs, ICCBs, and MCCBs must interrupt a fault within three cycles at maximum short-circuit current to protect the bus. This may necessitate using an instantaneous unit on the main circuit breaker to protect the bus at the sacrifice of obtaining selectivity.

The source or tie circuit breakers that carry current to the switchboard bus are set to provide long-time thermal overload and short-circuit protection for the bus, but the outgoing feeder circuit breakers act to provide protection against short-circuit currents that flow through the bus to faults external to the bus. Due to lack of short-time testing requirements, the outgoing feeder circuit breakers should be instantaneously tripped to meet the three-cycle testing time limit. For sensitive tripping of switchboard arcing phase-to-ground short circuits on solidly grounded systems of more than 150 V to ground, advantage should be taken of the enhanced protection offered by integral ground trips available on larger size MCCBs and on most LVPCBs. Switchboard arcing phase-to-ground short-circuit currents could be small with respect to main circuit breaker long-time pickup and, without the sensitive ground-trip protection, might remain undetected, causing severe damage. The ground-fault trips act fast and have low pickup currents, usually a fraction of the long-time pickup setting.

4.11.3 Protection of motor feeders and motors

4.11.3.1 Motor feeders

Motor feeders receive particular attention from NEC Article 430, which governs the selection of the current-carrying capacity of conductors used for motor applications. After the cable size is selected in accordance with this article, the overload, short circuit, and ground-fault protection are applied in accordance with NEC Articles 240 and 430. Motor overload protection is discussed in NEC Article 430, Part C. Short-circuit and ground-fault protection for both individual and grouped motor applications are discussed in NEC Article 430, Parts D and E. Maximum rating or settings of motor branch-circuit, short-circuit, and ground-fault devices is provided in NEC Table 430-152. Ground-fault trip units may be required on motor feeders to obtain coordination with ground-fault trip units on the supply devices.

NEC Article 430 requires the continuous-current rating of the circuit breaker to be no less than 115% of motor full-load amperes (FLA). The trip setting for an inverse time circuit breaker should not exceed 250% of motor full-load amperes (refer to NEC Table 430-152). Exceptions of 300% for over 100 A and 400% for under 100 A full-load ampere are permitted, if necessary for starting [refer to NEC Section 430-52, Exception 1(c).] It is desirable to use the lowest value of continuous-current rating, which ensures nuisance-free starting with its maximum instantaneous trip setting.

4.11.3.2 Motors

Selection considerations for motors are as follows:

a) Motor and branch-circuit overcurrent protection
b) Motor and branch-circuit short-circuit protection

See Figure 4-15 for an example of individual motor and branch-circuit protection. The 70 A circuit breaker is primarily intended to protect against short circuits. It should not generally be used as a switch to energize or de-energize motors. A contactor in a motor starter, rather than a circuit breaker, should be used for switching a frequently started motor on and off. For large motors that are started infrequently, LVPCBs may be applied (see Tables 4-5 and 4-6).

High-efficiency (NEMA Class E) motors, because of their low-loss designs, have higher starting current than standard NEMA motors of the same horsepower and code letters. It is not uncommon for certain high-efficiency motors to reach locked-rotor magnitudes of 800% of full-load current. If this is not taken into account in setting the breaker, then nuisance tripping will be the result. The circuit breaker's instantaneous trip unit must be 200% of locked-rotor current to avoid nuisance tripping when the motor is started.

Other potential application limitations that should be reviewed with the manufacturer are as follows:

— Frequent starting duty
— Extended starting times
— Ground fault coordination

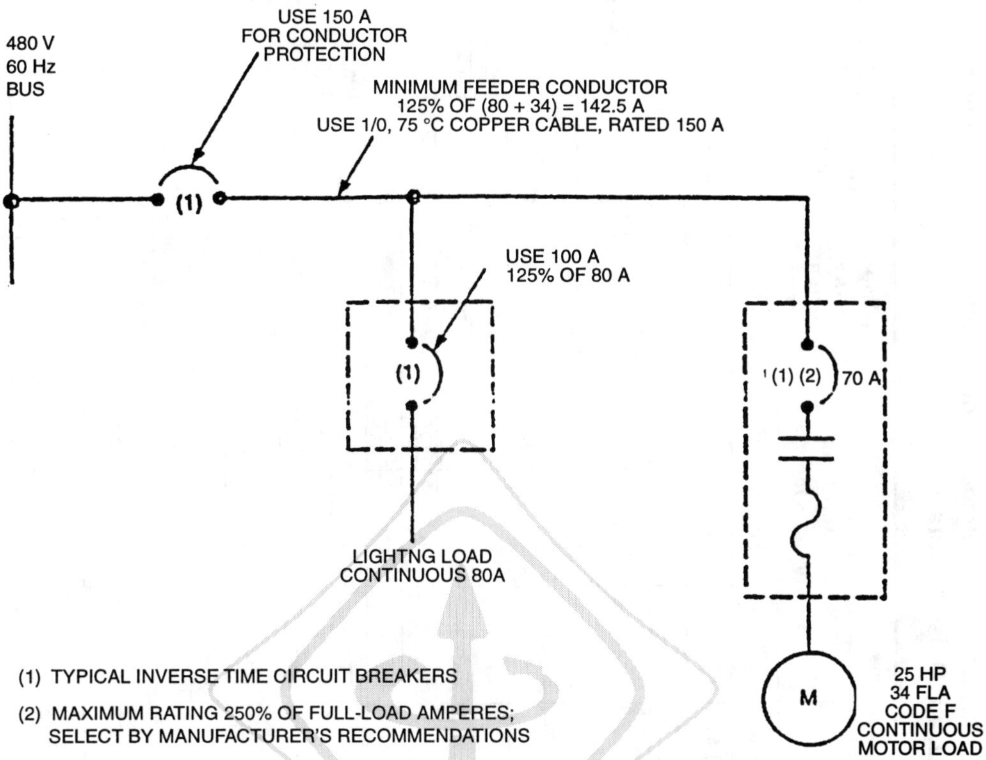

Figure 4-15 —Typical feeder circuit (lighting load and single fixed motor load)

Table 4-5—Application of low-voltage ac power circuit breakers to full-voltage motor starting and running duty of three-phase, 60 Hz, 40 °C rise motors[a]

Line no	Horsepower rating of three-phase ac motors									Trip device current rating (A)[b]	Motor full-load current (A)	
	Induction motors			100% power-factor synchronous motors			80% power-factor synchronous motors					
	230 V	460 V	575 V	220 V	440 V	550 V	220 V	440 V	550 V		Min	Max
	Col 1	Col 2	Col 3	Col 4	Col 5	Col 6	Col 7	Col 8	Col 9	Col 10	Col 11	Col 12
1	10	25	30	—	30	40	—	25	25; 30	40	26	35
2	15	30	40	—	40	50	—	30	40	50	32	44
3	20	40	50; 60	25	50	60	—	40	50	70	45	61
4	25; 30	50; 60	75	30	60	75	25	50	60	90	58	78
5	—	—	—	40	75	100	30	60	80	100	64	87
6	40	75	100	50	100	125	40	75	100	125	80	109
7	50	100	125	60	—	150	—	—	—	150	96	131
8	—	—	150	—	125	—	50	100	125	175	112	152
9	60	125	—	75	150	200	60	125	150	200	128	174
10	75	150	200	—	—	—	—	—	—	225	144	196
11	—	—	—	100	200	—	75	150	200	250	160	218
12	100	200	250[a]	—	—	—	—	—	—	300	192	261
13	—	250[a]	300	125	—	—	100	200	—	350	224	304
14	125	—	350	—	—	—	125	—	—	400	256	348
15	150	300; 350	400; 450	—	—	—	—	—	—	500	320	435
16	200	400	500	—	—	—	—	—	—	600	384	522
17	250[a]	450; 500	—	—	—	—	—	—	—	800	512	696
18	300; 350	—	—	—	—	—	—	—	—	1000	640	870
19	400	—	—	—	—	—	—	—	—	1200	768	1044
20	450; 500	—	—	—	—	—	—	—	—	1600	1023	1392

Table 4-5—Application of low-voltage ac power circuit breakers to full-voltage motor starting and running duty of three-phase, 60 Hz, 40 °C rise motors *(Continued)*

NOTES
1—*Locked-rotor current and instantaneous trip setting*. Circuit breakers selected from this table are suitable for all motors having locked-rotor kilovoltampere per horsepower, indicated by code letters A through J, inclusive, as listed in NEC Section 430-7. For motors with higher locked-rotor currents, care must be taken to ensure that an instantaneous trip setting high enough to permit motor starting is available. It may be necessary to choose the circuit breaker with the next higher continuous-current rating, provided that the calibration limitations given in Footnote b are not exceeded.
If motor locked-rotor current exceeds 600% of the circuit-breaker frame size, a shorter service life than that shown in ANSI C37.16-1988 can be expected.

2—*Applications to motors other than those listed.* For motors with horsepower ratings not listed in this table or for motors with other than normal speed or torque characteristics, it will be necessary to determine the full-load current and locked-rotor current as specified by the motor manufacturer. Find the current range in columns 11 and 12 that matches the full-load current to determine the circuit breaker with the proper continuous rating. Check locked-rotor current according to NOTE 1.

[a]Characteristics of motors of more than 200 hp vary widely, and the manufacturers of the motor should be consulted for specific details in these cases.
[b]*Selection of trip device current rating and circuit-breaker frame size.* The trip-device rating listed is a preferred rating from ANSI C37.16-1988. In accordance with NEC Section 430-110, this rating is at least 115% of the maximum motor full-load current (column 12). With trip devices having the lowest calibration point at 80% of the trip-device rating, the requirement of NEC Section 430-34 can be met for the minimum full-load current (column 11). NEC Section 430-34 requires that the trip device be set at a calibration point that does not exceed the following:

— 140% of motor full-load current for motors with a marked service factor not less than 1.15 and for motors with a marked temperature rise not over 40 °C.

— 130% of motor full-load current for all other motors.

Any value listed in column 10 may also be a trip-device setting if this current can be carried continuously and if additional adjustments allow compliance with NEC Section 430-34.

Trip devices having a higher current rating may be used provided that they have a suitable calibration point below 80% of the trip-device rating. The circuit-breaker frame size should be selected based on the applicable trip-device rating as well as the short-circuit current available. See Tables 1 and 2 of ANSI C37.16-1988 for guidance.

Source: ANSI C37.16-1988.

Table 4-6—Application of integrally fused low-voltage ac power circuit breakers to full-voltage motor starting and running duty of three-phase, 60 Hz, 40 °C rise motors—Maximum short-circuit current rating: 200 000 rms symmetrical current

Line no.	Horsepower rating of three-phase ac motor[a]									Trip device current rating (A)[b]	Typical rating of current limiting fuse (A)[c]	Motor full-load current (A)	
	Induction motors			100% power-factor synchronous motors			80% power-factor synchronous motors						
	230 V	460 V	575 V	220 V	440 V	550 V	220 V	440 V	550 V			Min	Max
	Col 1	Col 2	Col 3	Col 4	Col 5	Col 6	Col 7	Col 8	Col 9	Col 10	Col 11	Col 12	Col 13
1	40	75	100	50	100	125	40	75	100	125	400	80	109
2	50	100	125	60	—	150	—	—	—	150	600	96	131
3	—	—	150	—	125	—	50	100	125	175	600	112	152
4	60	125	—	75	150	200	60	125	150	200	600	128	174
5	75	150	200	—	—	—[a]	—	—	—	225	800	144	196
6	—	—	—[a]	100	200	—	75	150	200	250	800	160	218
7	100	200	—	—	—[a]	—	—	—	—[a]	300	1000	192	261
8	—	—[a]	—	125	—	—	100	200	—	350	1200	224	304
9	125	—	—	—	—	—	125	—[a]	—	400	1200	256	348
10	150	—	—	—	—	—	—	—	—	500	1600	320	435
11	200	—	—	—	—	—	—	—	—	600	2000	384	522

Table 4-6—Application of integrally fused low-voltage ac power circuit breakers to full-voltage motor starting and running duty of three-phase, 60 Hz, 40 °C rise motors— Maximum short-circuit current rating: 200 000 rms symmetrical current *(Continued)*

NOTES
1—*Locked-rotor current and instantaneous trip setting.* Circuit breakers selected from this table are suitable for all motors having locked-rotor kilovoltampere per horsepower, indicated by code letters A through J, inclusive, as listed in NEC Section 430-7. For motors with higher locked-rotor currents, care must be taken to ensure that an instantaneous trip setting high enough to permit motor starting is available. It may be necessary to choose the circuit breaker with the next higher continuous-current rating, provided that the calibration limitations given in Footnote b are not exceeded.

If motor locked-rotor current exceeds 600% of the circuit-breaker frame size, a shorter service life than that shown in ANSI C37.16-1988 can be expected.

2—*Applications to motors other than those listed.* For motors with horsepower ratings not listed in this table, or for motors with other than normal speed or torque characteristics, it will be necessary to determine the full-load current and locked-rotor current as specified by the motor manufacturer. Find the current range in columns 12 and 13 that matches the full-load current to determine the circuit breaker with the proper continuous rating. Check locked-rotor current according to NOTE 1.

[a]Characteristics of motors at more than 200 hp vary widely, and the manufacturers of the motor should be consulted for specific details in these cases.
[b]*Selection of trip-device current rating and circuit-breaker frame size.* The trip-device rating listed is a preferred rating from ANSI C37.16-1988. In accordance with NEC Section 430-110, this rating is at least 115% of the maximum motor full-load current (column 13). With trip devices having the lowest calibration point at 80% of the trip-device rating, the requirement of NEC Section 430-34 can be met for the minimum full-load current (column 12). NEC Section 430-34 requires that the trip device be set at a calibration point that does not exceed the following:

— 140% of motor full-load current for motors with a marked service factor not less than 1.15 and for motors with a marked temperature rise not over 40 °C.

— 130% of motor full-load current for all other motors.

Any value listed in column 10 may also be a trip-device setting if this current can be carried continuously and if additional adjustments allow compliance with NEC Section 430-34.

Trip devices having a higher current rating may be used provided that they have a suitable calibration point below 80% of the trip-device rating. The circuit-breaker frame size should be selected based on the applicable trip-device rating as well as the short-circuit current available. See ANSI C37.16-1988 for guidance.

[c]These ratings are based on the use of a direct-acting phase trip device with instantaneous trip element. Where information is available, the fuse rating may be selected to suit the particular application based on motor current, overcurrent trip characteristics, fuse melting time characteristics, and system coordination requirements.

Source: ANSI C37.16-1988.

4.11.3.3 Special-purpose circuit breakers

Combination motor starters contain instantaneous only circuit breakers. These are discussed in Chapter 6.

4.11.3.4 Examples of feeder and branch-circuit protection

Figures 4-16 and 4-17 are one-line diagrams of typical applications of circuit breakers used as mains and feeders. Figure 4-16 shows a typical MCC with a main MCCB and a motor starter with an instantaneous only circuit breaker. Figure 4-17 shows typical switchboard applications. The switchboard at the top has a main LVPCB and a feeder LVPCB. The lower switchboard has lugs only on the incoming cables and feeder MCCBs. One of the MCCBs is a feeder circuit breaker to an individual motor starter. Example setting calculations for these applications are given in 14.5.2 of IEEE Std 242-1986. Example coordination curves are given in Chapters 5 of this recommended practice and Chapter 14 of IEEE Std 242-1986.

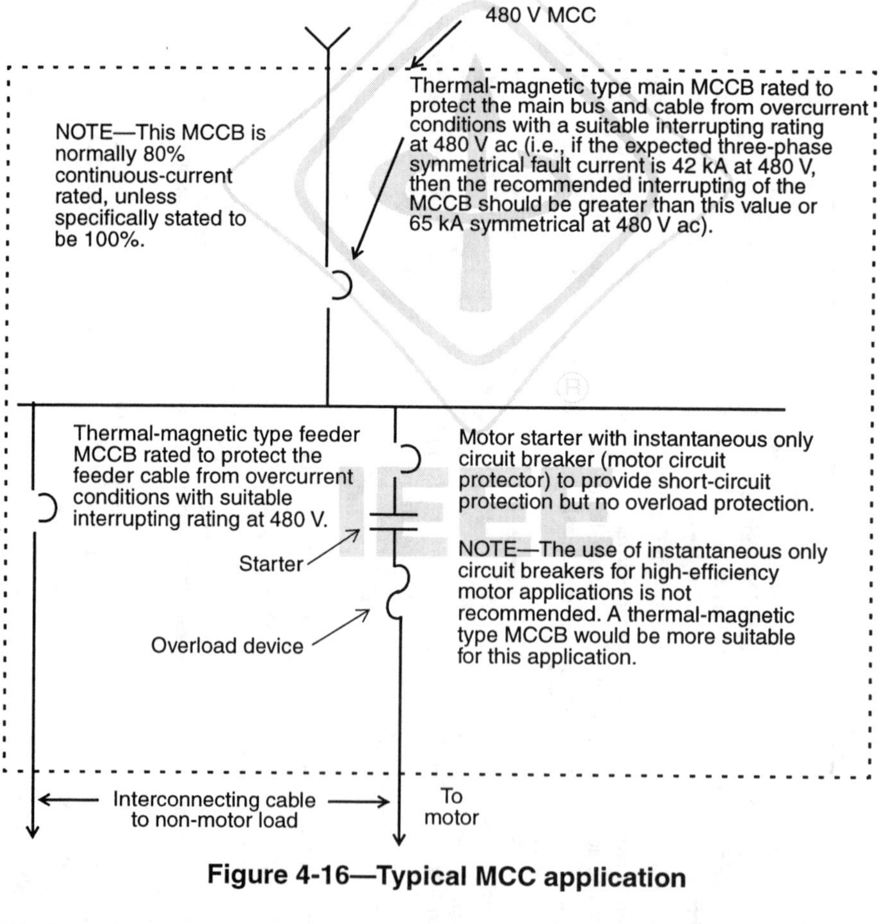

Figure 4-16—Typical MCC application

NOTE—It may be necessary to add a ground-fault trip function to the MCC feeders to obtain coordination with the ground-fault devices on the line-side devices, or if necessary, to specify shunt-trip devices for the feeder MCCBs.

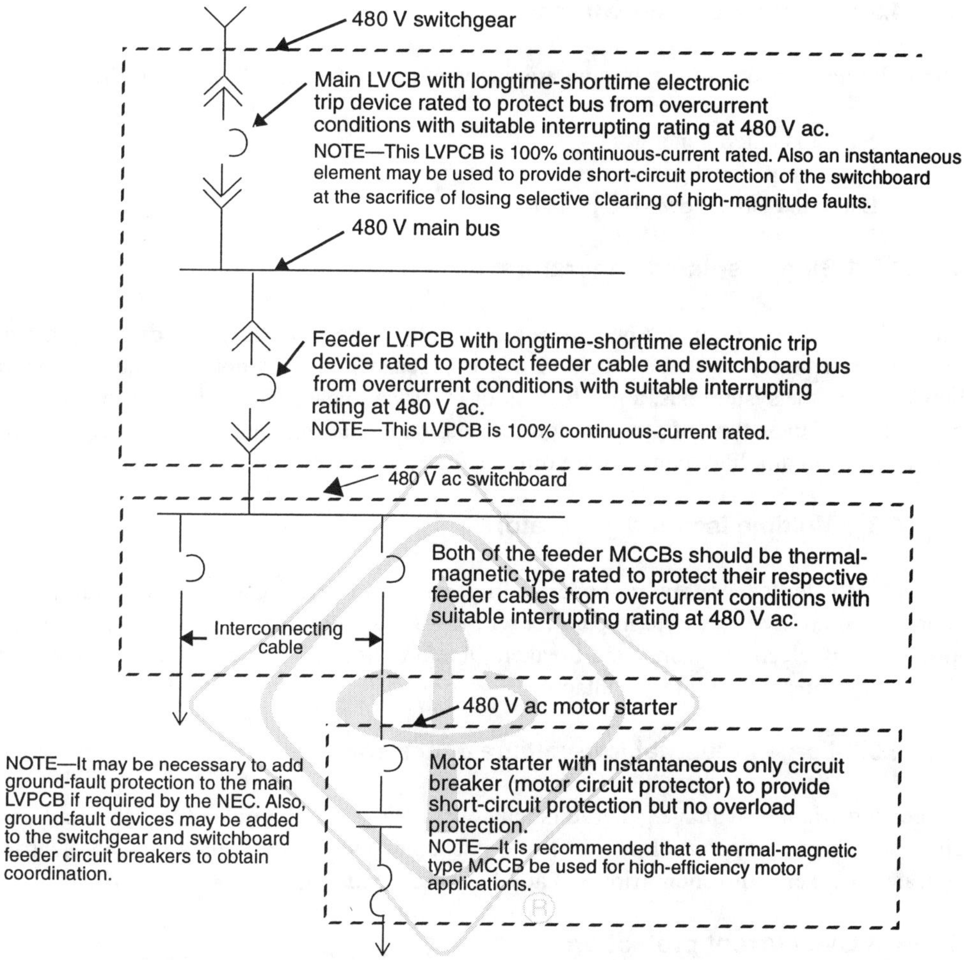

Figure 4-17—Typical switchgear/switchboard application

4.11.4 Protection of generators

4.11.4.1 Overcurrent protection

NEC Section 445-4 and Chapter 11 of IEEE Std 242-1986 discuss protection of generators.

4.11.4.2 Application considerations

Considerations for applying circuit breakers for generator applications should include the following:

a) Isolated or parallel operation
b) Short-circuit duty
c) Overload and extended short-circuit capability
d) Generator ratings
e) Type of grounding

4.11.4.3 Generator classifications

The following classifications of generators are defined in Chapter 11 of IEEE Std 242-1986:

a) Single isolated generators
b) Multiple isolated generators
c) Large industrial low-voltage generators

4.11.4.3.1 Single isolated generators

Single isolated generators do not operate in parallel with the utility source. They are often used for emergency or standby power. If a transfer switch does not separate the generator breaker from the system bus, a mechanical or electrical means should be provided to prevent paralleling. If their type of service requires automatic starting and shutdown, electric close, shunt trips, and auxiliary contacts are often necessary.

4.11.4.3.2 Multiple isolated generators

Multiple isolated generators may operate in parallel with each other, but not with the utility source. The ability to operate in parallel will involve synchronizing circuits which require a quick-closing electrically operated breaker. Other control and protection devices may necessitate shunt trips and auxiliary contacts.

4.11.4.3.3 Large industrial low-voltage generators

Large industrial low-voltage generators may operate in parallel with a utility source. For synchronous generators, control and protection schemes will require quick-closing electrically operated breakers with shunt trips and auxiliary contacts.

4.11.4.4 Overcurrent protection

Circuit-breaker trip devices for generator circuits should generally include the following characteristics:

a) Long-time for protection against prolonged low-level overload conditions.

b) Short-time for selectivity with feeder breakers to protect against bus faults and system faults not cleared by feeder breakers.

c) Instantaneous to trip on faults in the generator or leads that are being fed from elsewhere in the system if the generator is paralleled with the utility system.

d) Ground-fault protection, which is required by the NEC in some specific applications.

4.11.4.5 Short-circuit considerations

The short-circuit rating of the generator circuit breaker should be equal to or greater than the short-circuit current available to it from the system, which is generally higher than that available from the generator.

In addition, most generators have an extremely rapid decrement of short-circuit current to almost zero. If the circuit breaker is to be tripped due to short circuit, the time-current characteristic of the protective device must be below the generator's short-circuit decrement curve. For detailed protection considerations, refer to Chapter 11 of IEEE Std 242-1986.

4.11.4.6 Additional protection considerations

The electronic trip device on circuit breakers is sufficient protection for many small generators. Larger, more costly generators may require additional protection as outlined in Chapter 11 of IEEE Std 242-1986.

If additional protective or control devices are desired, appropriate accessories should be included on the circuit breaker.

a) Protective relays will require a shunt trip on the circuit breaker. This, in turn, will necessitate a reliable source of control power or a capacitor trip device.

b) Automatic startup will require electrical operators.

c) Synchronizing will require a quick-closing electrically operated circuit breaker.

4.11.5 Protection of capacitors

A circuit breaker can be used to provide the overcurrent protection for low-voltage capacitor systems as required by NEC Section 460-8. The circuit breaker should have a voltage rating suitable for the rated voltage of the capacitor system. In addition, the circuit breaker must have an interrupting rating greater than the fault current available at its line-side terminals. However, the NEC, in the same section, only requires that "the rating or setting of the overcurrent device shall be as low as practicable." Therefore, selection of the overcurrent rating of a circuit breaker for protection of capacitor systems needs to be given further consideration.

4.11.5.1 Application considerations

Considerations in the application of low-voltage breakers for unit capacitor supply should include the following:

a) Transient inrush current for isolated bank or parallel bank switching configurations
b) Transient overvoltages generated during opening operations by restrikes
c) Protective device characteristics
d) Frequent switching requirements
e) High-frequency inrush currents
f) Continuous-current requirements

NOTE—Parallel bank fault duty contribution has traditionally been ignored based on short persistence.

4.11.5.2 Conductor and protective device sizing

The capacitor circuit conductors and disconnecting means are required to have an ampacity not less than 135% of the rated current of the capacitor per NEC Section 460-8. Capacitors are generally manufactured with a 15% tolerance, so that a 100 kvar capacitor may actually draw a current equivalent to a 115 kvar capacitor. In addition, the current drawn by a capacitor varies directly with the line voltage. A variation in the line voltage from a pure sine wave, such as during a switching transient, causes the capacitor to draw an increased current. Also, system harmonics will be a contributing factor to increased current. Considering these factors, the actual current in an installed capacitor system can amount to 135% of the rated current of the capacitor. A typical circuit breaker and capacitor circuit are shown in Figure 4-18.

As with the conductors and disconnecting means, the overcurrent protective device should also be rated for at least 135% of the capacitor bank current rating. However, it must be rated low enough to protect the capacitor bank from violent damage. Due to the variations mentioned above, manufacturers generally recommend values over 150% of the capacitor bank current ratings (typically in the 165–200% range). It is suggested that the specific application be discussed with the manufacturers of the capacitors and circuit breakers before deciding on the rating or setting of the circuit breaker. To achieve improved protection, capacitor banks with multiple cans per phase may have the individual cans fused.

In addition, verify that the circuit breaker is capable of switching capacitor current of 180 A.

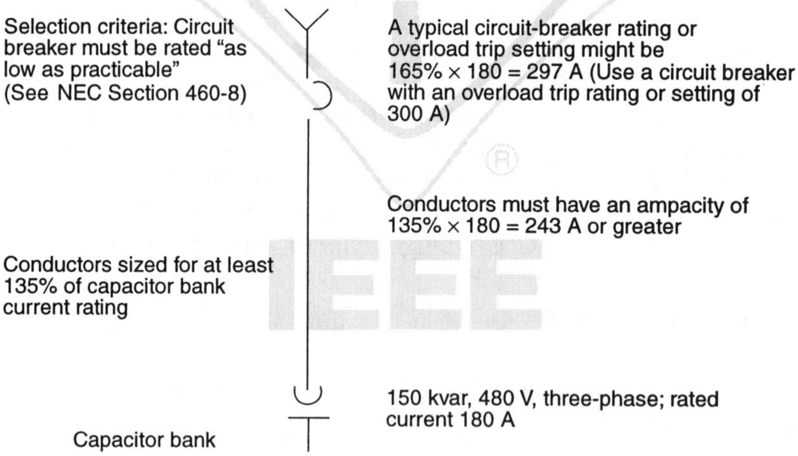

Figure 4-18—Capacitor protection

4.11.5.3 Ambient temperature considerations

For application in ambients higher than the rated ambient of the circuit breaker, the manufacturer should be consulted to determine the rating of the circuit breaker required to meet the minimum of 135% capacitor rating. In locations where temperatures vary greatly, ambient compensating circuit breakers may be desirable.

4.11.6 Protection of transformers

Low-voltage circuit breakers can be used for the transformer protection required by NEC Section 450-3. Refer to this section for the specific primary and secondary protection requirements for transformers over 600 V nominal and equal to or less than 600 V nominal. The protection discussed in this section of the NEC is intended to protect the transformer only. Protection of the primary and secondary conductors may be obtained by proper selection of cables.

4.11.6.1 Application considerations

Considerations for the application of low-voltage circuit breakers for transformer protection include the following:

a) Will they clear the system for short circuits within the transformer?

b) Will they prevent the transformer from becoming overloaded beyond its ability?

c) Will they protect the transformer from damage during a through-fault condition on the load side?

d) Do they have adequate interrupting ratings for faults at their load-side terminals?

e) Will they handle the transformer inrush current without nuisance tripping?

f) Can they tolerate the current transients during inrush and during other operating conditions?

g) Do they provide conductor protection?

h) Is ground-fault protection provided (if required)?

4.11.6.2 Transformer with a primary rated over 600 V

When the transformer primary is over 600 V and the secondary is 600 V or less, low-voltage circuit breakers might be used as the secondary transformer protection. The rating of this secondary protection must not exceed 125% of the transformer rated secondary current, or the next higher standard rating or setting for unsupervised transformer applications per NEC Section 450-3(a)(1); NEC Section 450-3(a)(2) allows 250% of the transformer rated secondary current for "supervised" installations.

As shown in Figure 4-19, the secondary circuit breaker's overload trip rating or setting must be below 751 A, or the next higher standard rating or setting. Refer to NEC Section 450-3(a)(1).

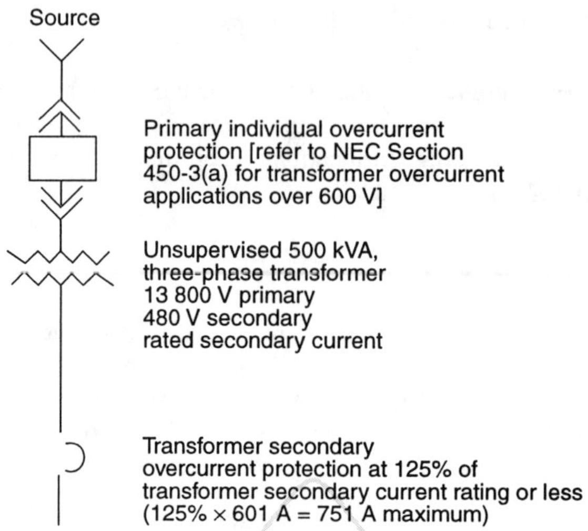

Figure 4-19— Transformer with primary and secondary protection

4.11.6.3 Transformer primary and secondary rated 600 V or below

4.11.6.3.1 Primary protection only

The overload ratings or settings determined by the following paragraph do not necessarily provide conductor protection. For example, NEC Section 240-3(i) states that transformer secondary conductors (other than two-wire) are not considered to be protected by the primary overcurrent protection. Before making the final selection of the circuit-breaker rating, conductor protection must be verified.

NEC Section 450-3(b)(1) states that if only primary protection is to be used for a transformer of 600 V or less, that protection shall be an individual overcurrent device on the primary side, rated or set at not more than 125% of the rated primary current of the transformer as shown in Figure 4-20. If the primary current rating of the transformer is less than 9 A, the exceptions allow the overcurrent device to be rated up to, but no more than, 167% of the transformer primary current rating. If the primary current rating of the transformer is less than 2 A, the exceptions allow the overcurrent device to be rated up to, but no more than, 300% of the transformer primary current rating.

4.11.6.3.2 Transformers with secondary protection

When the transformer has secondary protection, an individual overcurrent device is not required on the primary side if

a) The overcurrent device on the secondary side is rated or set at not more than 125% of the transformer secondary rating, and

b) The primary feeder overcurrent device is rated or set at not more than 250% of the transformer primary current rating [refer to NEC Section 450-3 (b)(2)].

An example of this protection is shown in Figure 4-21.

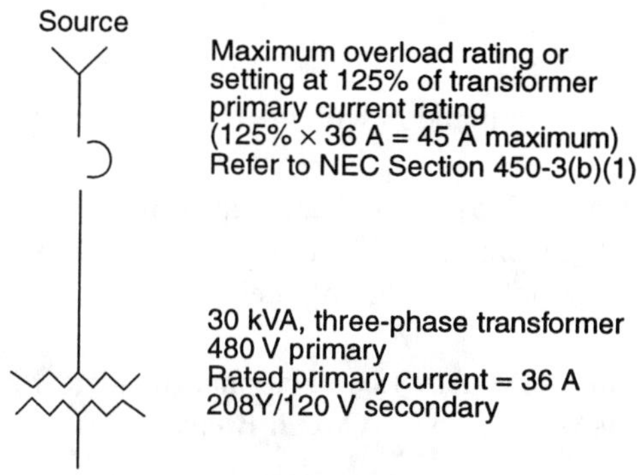

Figure 4-20—Individual circuit breaker in transformer primary for an "unsupervised" transformer

The NEC guidelines provide the maximum circuit breaker ratings/settings; lower ratings/settings are recommended for improved protection. Protective devices on the transformer primary and secondary circuits are recommended for the best transformer protection.

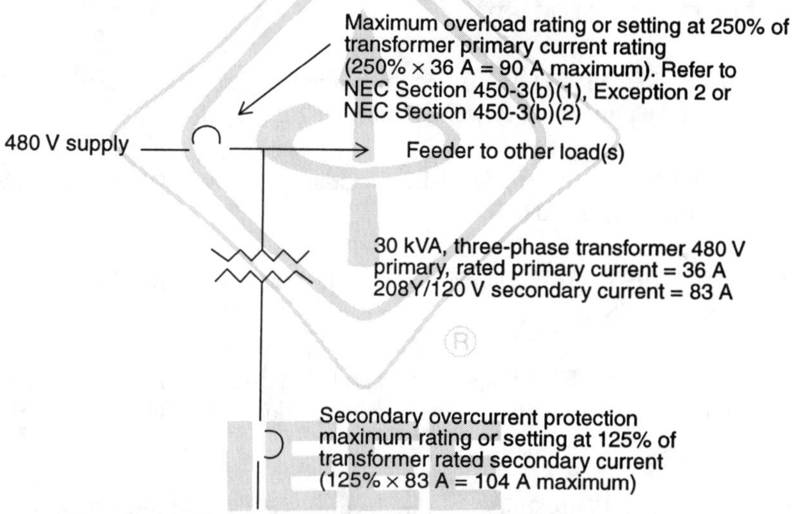

Figure 4-21—Circuit breaker protecting feeder supplying a transformer and other load(s)

4.11.6.4 Other considerations for protecting transformers

Selecting the current ratings is only part of the job of protecting the transformer. The answers to some of the questions in 4.11.2 involve considerations of time as well as current. Transformer damage curves, current inrush data, overload capabilities, and information on transient tolerances can be obtained from the manufacturers of the transformers and IEEE standards. Refer to the IEEE C57 Collection and IEEE Std 242-1986. This type of information will help the designer determine the proper trip unit settings. Selective coordination examples are shown in Chapter 5. To obtain reliable operation, proper acceptance testing and maintenance of circuit breakers is required (refer to Chapter 7).

4.12 References

This chapter shall be used in conjunction with the following publications:

ANSI C37.16-1988 (Reaff 1995), American National Standard for Switchgear—Low-Voltage Power Circuit Breakers and AC Power Circuit Protectors—Preferred Ratings, Related Requirements, and Application Recommendations.[3]

ANSI C37.17-1979 (Reaff 1988), American National Standard for Trip Devices for AC and General Purpose DC Low-Voltage Power Circuit Breakers.

ICEA P-32-382-1994, Short-Circuit Characteristics of Insulated Cable.[4]

IEEE Std 141-1993, IEEE Recommended Practice for Electric Power Distribution for Industrial Plants *(IEEE Red Book)* (ANSI).[5]

IEEE Std 241-1990, IEEE Recommended Practice for Electric Power Systems in Commercial Buildings *(IEEE Gray Book)* (ANSI).

IEEE Std 242-1986 (Reaff 1991), IEEE Recommended Practice for Protection and Coordination of Industrial and Commercial Power Systems (*IEEE Buff Book*) (ANSI).

IEEE Std C37.13-1990 (Reaff 1995), IEEE Standard for Low-Voltage AC Power Circuit Breakers Used in Enclosures (ANSI).

IEEE Std C37.20.1-1993, IEEE Standard for Metal-Enclosed Low-Voltage Power Circuit Breaker Switchgear (ANSI).

IEEE Std C57.12.01-1989, IEEE Standard General Requirements for Dry-Type Distribution and Power Transformers Including Those with Solid Cast and/or Resin-Encapsulated Windings.

IEEE Std C57.109-1993, IEEE Guide for Liquid-Immersed Transformer Through-Fault-Current Duration (ANSI).

IEEE Distribution, Power, and Regulating Transformers Standards Collection (C57), 1995 Edition.

NEMA AB 3-1996, Molded Case Circuit Breakers and Their Application.[6]

[3]ANSI publications are available from the Sales Department, American National Standards Institute, 11 West 42nd Street, 13th Floor, New York, NY 10036, USA.

[4]ICEA publications are available from ICEA, P.O. Box 411, South Yarmouth, MA 02664, USA.

[5]IEEE publications are available from the Institute of Electrical and Electronics Engineers, 445 Hoes Lane, P.O. Box 1331, Piscataway, NJ 08855-1331, USA.

[6]NEMA publications are available from the National Electrical Manufacturers Association, 1300 N. 17th St., Ste. 1847, Rosslyn, VA 22209, USA.

NEMA PB 2-1995, Deadfront Distribution Switchboards.

NFPA 70-1996, National Electric Code® (NEC®).[7]

UL 489-1991, Molded-Case Circuit Breakers and Circuit-Breaker Enclosures (DoD).[8]

UL 891-1994, Dead-Front Switchboards (DoD).

4.13 Bibliography

Additional information may be found in the following source.

[B1] UL Recognized Component Directory, vol. 1, 1996.

[7]The NEC is available from Publications Sales, National Fire Protection Association, 1 Batterymarch Park, P.O. Box 9101, Quincy, MA 02269-9101, USA. It is also available from the Institute of Electrical and Electronics Engineers, 445 Hoes Lane, P.O. Box 1331, Piscataway, NJ 08855-1331, USA.

[8]UL standards are available from Global Engineering, 1990 M Street NW, Suite 400, Washington, DC, 20036, USA.

IEEE

Chapter 5
Selective coordination of low-voltage circuit breakers with other protective devices

5.1 Introduction

A selectively coordinated power system has protective devices that isolate the smallest portion of the system when interrupting a short circuit or overload and thus limit damage to components. This is accomplished with low-voltage circuit breakers by the selection of appropriate operating ratings, trip characteristics, and trip settings so that only the closest circuit breaker on the source side of an overcurrent condition clears the abnormality.

5.1.1 Time-current curves

The low-voltage circuit breaker has a protective element that operates in response to the magnitude and duration of current passing through it. It is direct acting in that the current through it provides energy to release the opening mechanism. The characteristic time-current curve has a band of operating area. The upper limit of the band represents the maximum total clearing time for the circuit breaker. The lower limit of the band shows the maximum resettable delay, i.e., the maximum time that a given amount of through current (e.g., a fault or overload) may persist and then subside without tripping the circuit breaker.

Bands indicating the pickup current (vertical asymptote) of the characteristic show the tolerance of the pickup point. Currents less than the long-time pickup band can be sustained without tripping the circuit breaker. Currents at or above the upper limit of the band will result in tripping of the circuit breaker.

5.2 LVPCBs

The low-voltage power circuit breaker (LVPCB) may be found in two general varieties, those with electromechanical trip devices, and those with electronic trip devices. Each has some combination of long-time delay, short-time delay, instantaneous and ground-fault trip elements. ANSI C37.17-1979[1] defines those characteristics and their limits. Dual trip devices have long-time and instantaneous elements. Selective trip devices have long-time and short-time elements. Triple selective devices have long-time, short-time, and instantaneous elements.

5.2.1 Electromechanical trip devices

The electromechanical trip (refer to Figure 5-1) uses a magnetic circuit directly applied to the circuit-breaker current-carrying conductor. A variety of springs, dashpots, and escapements provide the time-current characteristic. The time delay band is relatively broad due to the tolerance of the mechanical devices, temperature, wear, age, etc. Sensitive ground-fault protection must be provided by external sensing and remote tripping devices. Those remote devices must trip the LVPCB by a shunt trip.

[1]Information on references can be found in 5.6.

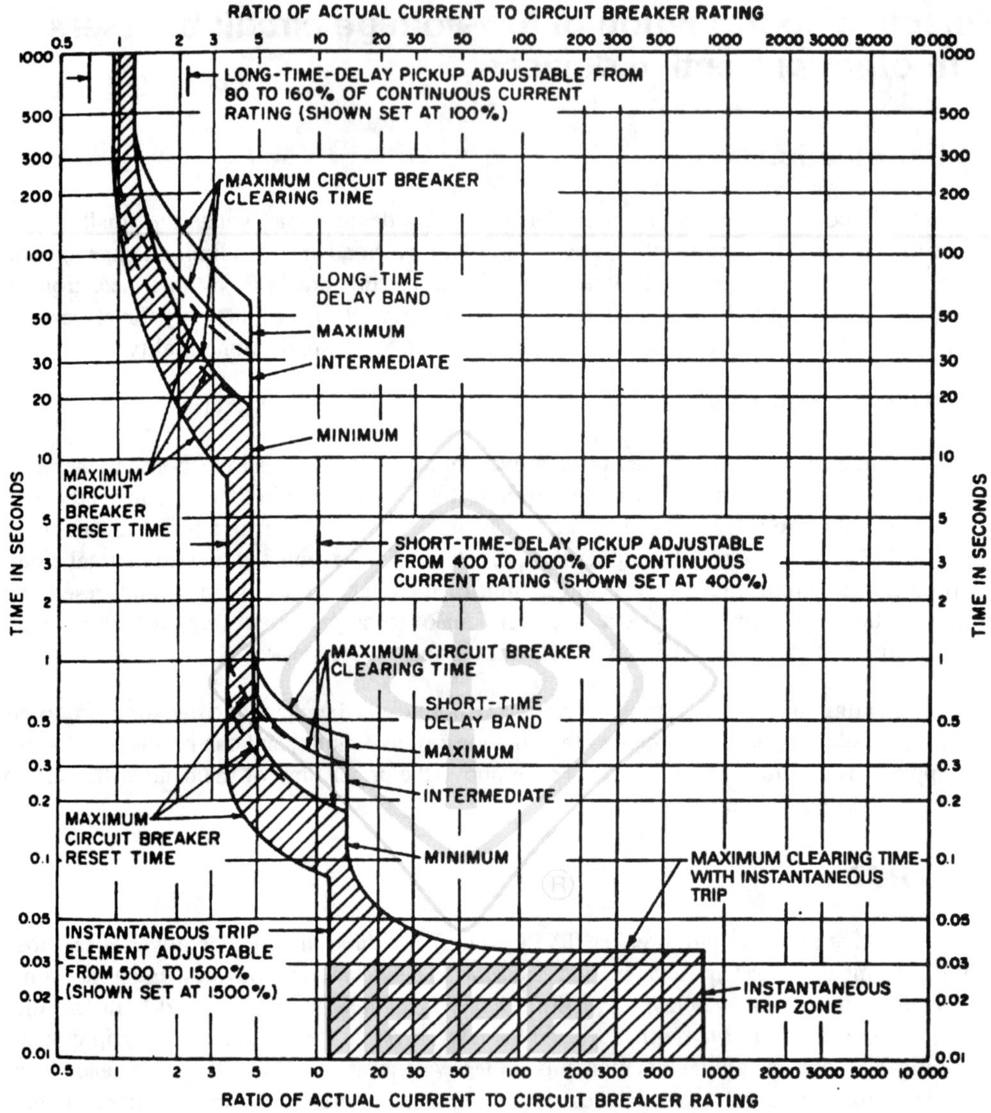

Figure 5-1—Typical time-current plot for electromechanical trip devices

5.2.2 Electronic trip devices

More recently, the electronic trip device (refer to Figure 5-2) has become the industry standard. Sensors detect the current and provide tripping energy. The time-current characteristic is electronically developed and is more accurate with narrower tolerances than for the electromechanical device. The long-time element provides overload protection and the short-time and instantaneous elements provide fault protection. Their characteristics are described in 5.2.2.1 through 5.2.2.4.

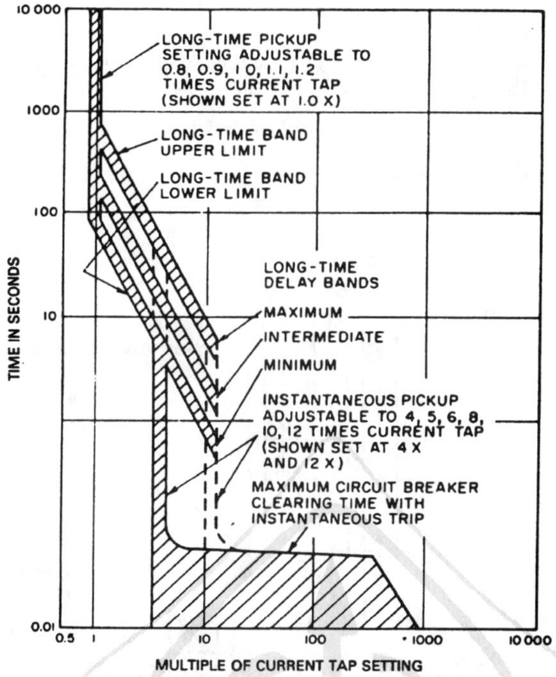

Figure 5-2—Typical time-current plot for electronic trip devices

5.2.2.1 Long-time delay protection

The pickup value of the long-time delay element is ±10% of the selected current. Currents greater than 90% of the pickup setting may eventually result in circuit-breaker tripping. The long-time delay is a constant I^2t inverse time characteristic to simulate the conductor and load equipment heating. Its pickup should be set for conductor and equipment overload protection as called for in the National Electrical Code® (NEC®) (NFPA 70-1996). Its delay should be set as low as possible, but coordinated with load-side devices. With some manufacturers' trip devices, adjustment of the long-time pickup setting will shift the long-time delay curve left or right; with others, the curve extends as the pickup setting is decreased, or it is chopped off as the pickup is increased. The pickup setting is usually some multiple of the unit's sensor rating.

5.2.2.2 Short-time delay protection

The short-time delay element provides a definite time delay. Adjustment allows selectivity with load-side instantaneous or faster short-time trip elements. Some devices offer an I^2t pickup, which provides an inverse characteristic between the pickup current and the fixed time delay (see Figure 5-3). This may provide an opportunity to better coordinate with load-side fuses. The short-time and instantaneous pickup currents are set above normal operating conditions (i.e., motor starting, transformer inrush, etc.).

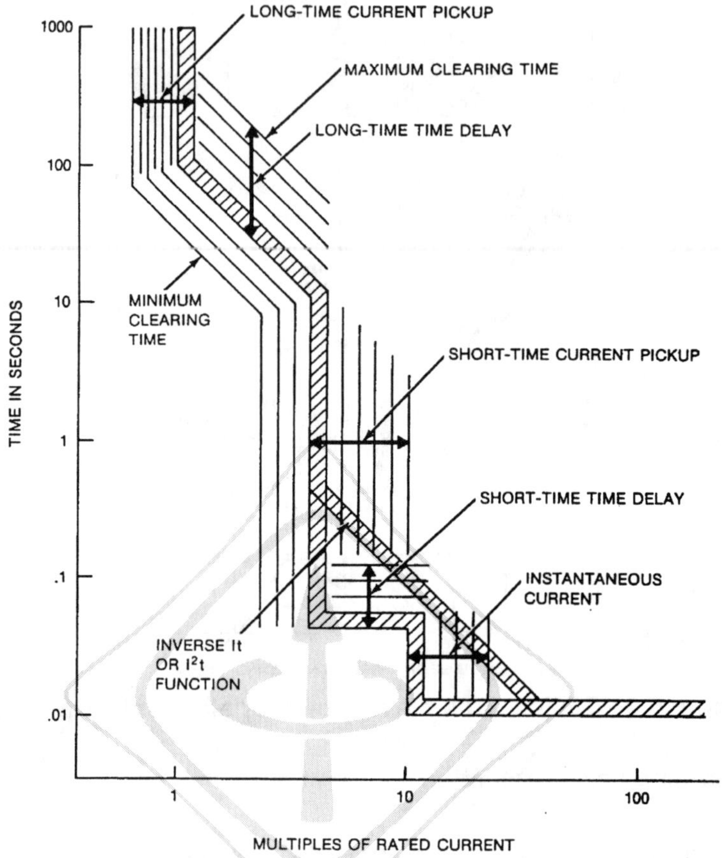

Figure 5-3—Time-current curve with short-time I^2t pickup

5.2.2.3 Instantaneous protection

An instantaneous trip has historically meant "no intentional delay." However, the electronic trip unit is desensitized near its pickup value to avoid nuisance tripping. This results in an "inverse instantaneous" pickup. At least one manufacturer employs a fault closing discriminator. In this instance, a circuit breaker closing on a fault may trip instantaneously even though there is no instantaneous pickup setting on the trip device.

5.2.2.4 Ground-fault protection

Sensitive ground-fault protection is accomplished using the sum of the three-phase currents, and adding the neutral current of four-wire systems. The resultant is the ground-fault current. Other systems use ground sensors that enclose all three-phase conductors, and the neutral of four-wire systems, to sense ground current. Still others sense the current flowing between the grounding electrode and the transformer neutral. Pickup currents are limited to 1200 A maximum to be consistent with the NEC. They are normally set more sensitive than the phase overcurrent protection since they do not respond to phase current. The time delay settings are similar to the short-time delay settings. As with short-time delay, the ground-fault trip characteristic may have the option of I^2t pickup.

5.3 Low-voltage MCCBs and ICCBs

As with the LVPCBs, low-voltage molded-case circuit breakers (MCCBs) are found in two general varieties: those with thermal-magnetic trip devices and those with solid-state trip devices. The insulated-case circuit breaker (ICCB) will have an electronic trip device. UL 489-1991 defines the requirements for both MCCBs and ICCBs.

5.3.1 Thermal-magnetic circuit breakers

The thermal-magnetic circuit breaker (refer to Figure 5-4) uses a bimetal or other similar device to establish its long-time delay characteristic and a spring and magnetic circuit for its instantaneous element. The long-time pickup is generally not adjustable. Sensitive ground-fault protection must be provided by auxiliary add-on or external devices, some using a shunt trip to open the circuit breaker. This shunt trip must be specified at the time of purchase.

5.3.2 Electronic trip devices

An electronic trip device is available on certain larger MCCBs and ICCBs. Some of these larger circuit breakers also have limited short-time delay capability.

5.3.3 Long-time and short-time delay protection

For either the thermal magnetic or solid-state trip device, the long-time pickup tolerance is –0% and not more than +25%. The circuit breaker will carry its pickup setting current without tripping. At 125% current, the circuit breaker will ultimately trip.

The long-time delay of the thermal-magnetic device is a good simulation of device heating, provided that the ambient temperatures are similar. The electronic long-time delay unit has an adjustable time, I^2t characteristic. MCCBs and ICCBs may have a short-time characteristic that gives a definite time delay, up to the instantaneous pickup current.

5.3.4 Instantaneous protection

Electronic devices have one of several types of instantaneous trips. Pickup current might be adjustable, or non-adjustable at a high-pickup current, or there may be a fault closing discriminator (see 5.2.2.3). Even though there is no indication of an instantaneous trip on the trip device, some devices instantaneously trip under heavy fault conditions.

5.3.5 Ground-fault protection

Ground-fault protection is available with the electronic trip device. The operating characteristic may be definite time, with or without an I^2t pickup.

Enhanced ground-fault protection for some equipment can be achieved by zone-selective interlocking. Typically, as shown in Figure 5-5, a main circuit breaker will be interconnected with each of its feeders. The main circuit breaker ground-fault trip device will operate in either of two modes. It will trip in a fast time if a ground fault is sensed in the main bus zone. For a ground fault on a feeder, the feeder circuit breaker will send a blocking signal to the main circuit breaker ground-fault trip, transferring it to a slow trip mode, allowing the feeder circuit breaker to clear the fault. Should the feeder fail to clear the fault, the delayed-trip main circuit breaker provides backup protection.

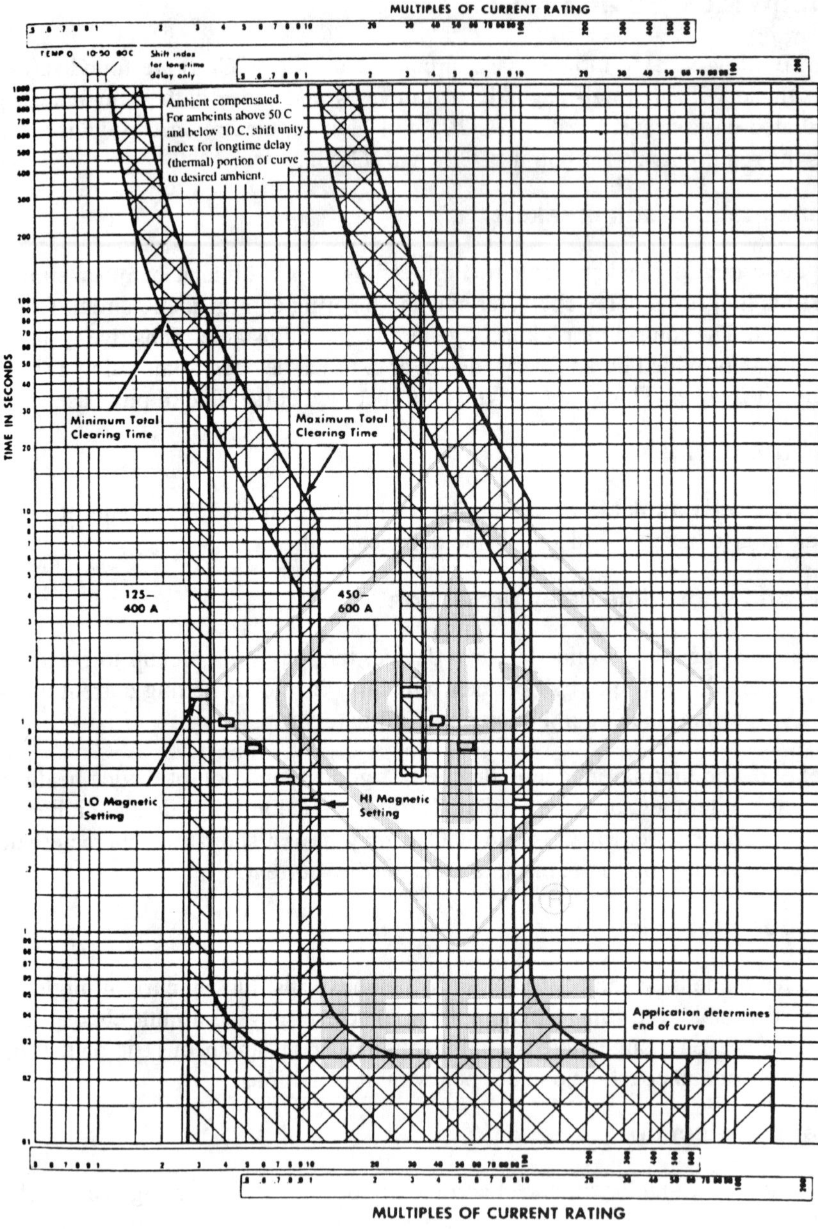

Figure 5-4—Time-current curves for 125–600 A MCCBs

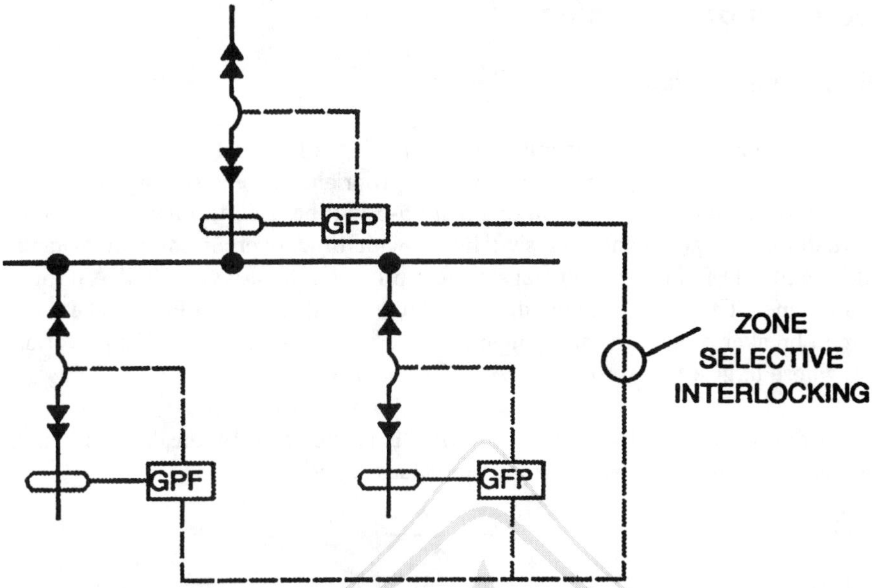

Figure 5-5—Zone-selective interlocking

5.4 Other coordinating devices

5.4.1 Low-voltage fuses

Low-voltage fuses, whether separate or part of a low-voltage circuit breaker, have a time-current band between two curves—the total clearing time curve (including the melting time tolerance and arcing time), and the minimum melting time curve.

5.4.2 Medium-voltage fuses

Medium-voltage fuses also have a time-current band between two curves. The total clearing time curve shows the maximum time from initiation of fault to total clearing of the circuit. The minimum melting time curve shows through currents that can be sustained without damage to the element. The minimum melting time curve may not show the effects of pre-loading currents.

5.4.3 Overcurrent relays

The operating characteristic of an overcurrent relay will be shown on the time-current curve as a line. The actual current interrupting time will depend on the reaction time of other components. Positive tolerance factors include auxiliary relay operating time, circuit breaker operating time, and current transformer saturation. Negative tolerance factors include disk over travel and current offset.

5.5 Coordination examples

5.5.1 Single-line diagram

To demonstrate coordination, the operating characteristics of several series protective devices may be plotted on log-log graph paper with appropriate current scaling among various voltage levels. The single-line diagram of Figure 5-6 will be used to illustrate a coordinated system. Medium-voltage metal-clad switchgear with time overcurrent and instantaneous relays (device 50/51) fed by current transformers protect a delta-wye 750 kVA 4160–480Y/277 V transformer. The 480 V main and feeder circuit breakers are LVPCBs. The highest set feeder circuit breaker protects a cable run terminated in a low-voltage MCCB power distribution panel whose largest breaker is a 100 A "E"-frame MCCB.

The time current curve of the 100 A frame MCCB is shown in Figure 5-7. This is a fixed curve; neither long-time nor instantaneous trip values are adjustable.

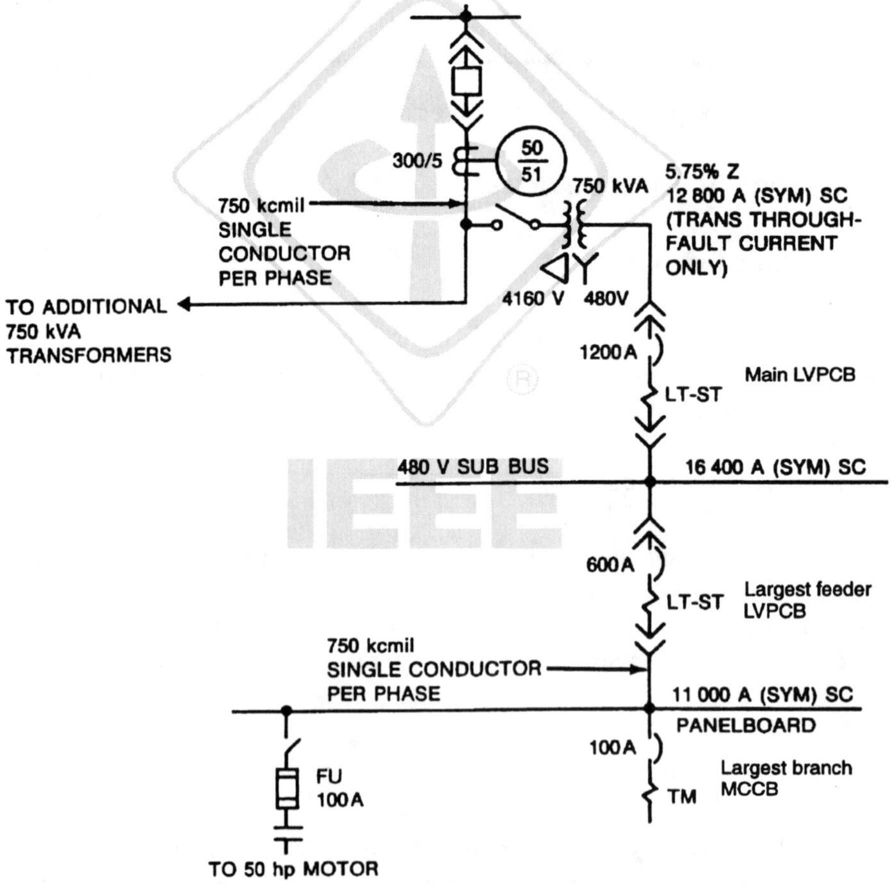

Figure 5-6—One-line diagram for coordination study

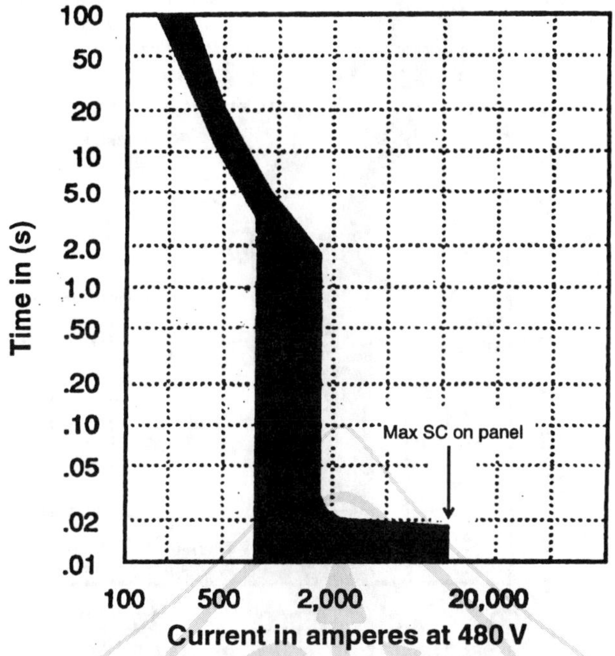

Figure 5-7—100 A frame circuit breaker

5.5.1.1 MCCBs and LVPCBc

Coordination of the 100 A MCCB and the 600 A feeder LVPCB is shown in Figure 5-8. The feeder power circuit breaker trip device will have adjustable settings of long-time pickup, long-time delay, short-time pickup, and short-time delay. An instantaneous trip element is not used on the feeder as it is not selective with the instantaneous trip element of the load-side 100 A MCCB.

The NEC requires that the long-time pickup of the 600 A circuit breaker protect for the ampacity of the 750 kcmil conductor, 475 A, so it is set for 80% of the 600 A rating (480 A). The long-time delay is selected with a longer time delay than the delay of the MCCB. The short-time pickup is selected to be higher than the 100 A circuit breaker instantaneous pickup tolerance. The short-time delay is set with more delay than clearing time of the MCCB instantaneous trip. A distribution panel fault tripped by the MCCB instantaneous element will not trip the feeder.

5.5.1.2 Main circuit breaker

The 1200 A main LVPCB also has adjustable long-time pickup, long-time delay, short-time pickup, and short-time delay settings. A 480 V main LVPCB will typically have its long-time pickup set at 125% of the transformer full-load current. Its remaining settings provide moderately longer operating times than for the largest feeder, as shown in Figure 5-9.

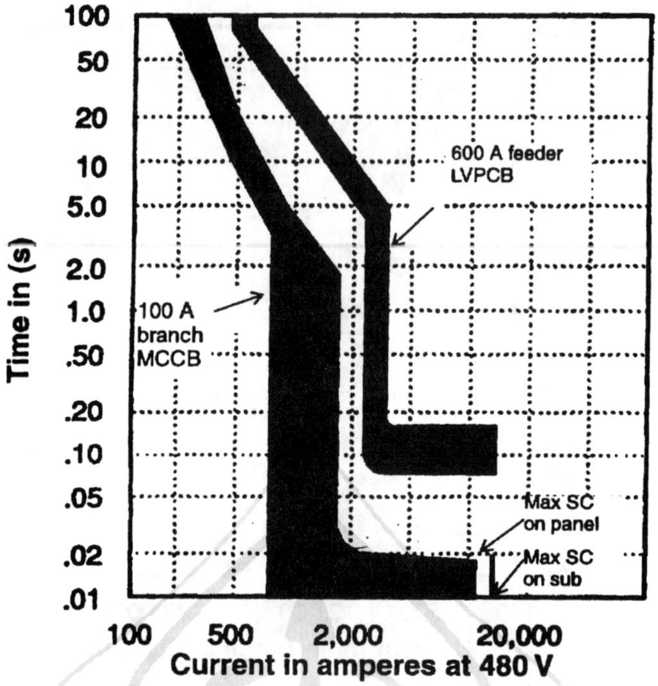

Figure 5-8—100 A frame and 600 A feeder circuit breaker

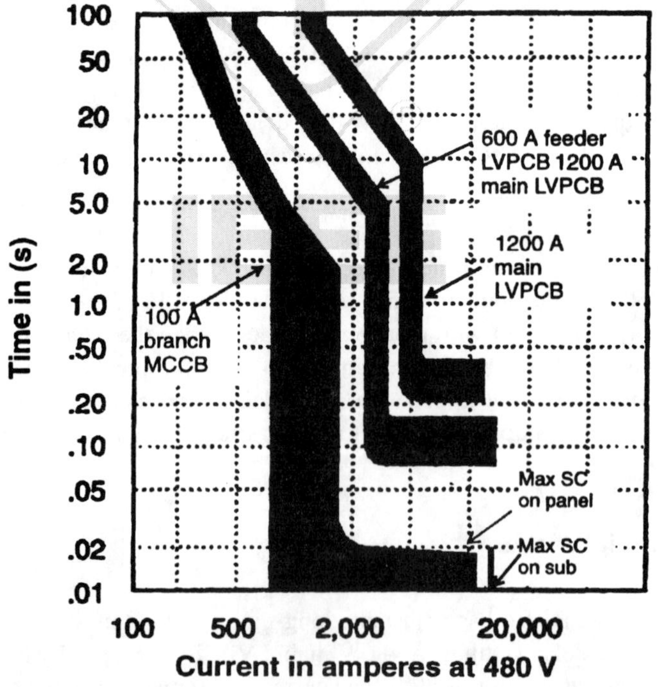

Figure 5-9—100 A frame, 600 A feeder, and 1200 A main circuit breakers

5.5.1.3 Primary overcurrent relay (device 50/51)

The transformer primary relay has to be set in accordance with a number of criteria, coordination among them. The NEC requires that its pickup must not exceed six times the transformer full-load current (this assumes that the transformer impedance is less than 6%). Its inverse time delay must protect the transformer for its through-fault withstand and yet should provide sufficient delay to exceed the time delay tripping of the low-voltage main circuit breaker. This setting should protect for the three-phase withstand rating. For delta-wye solidly grounded transformers, it should also protect for 57% of three-phase currents for line-to-ground through-faults (87% for ungrounded, or high-resistance grounded secondaries). The instantaneous element will be set higher than the transformer inrush (approximately 8–12 times full load for 0.1 s). It is also set high enough so that it does not trip for 480 V system faults. Figure 5-10 shows the coordinated elements for the entire system.

The primary protective device on a delta-wye transformer has a further consideration in that for a secondary phase-to-phase fault, one primary circuit will see 16% more current than a three-phase fault of the same secondary current magnitude. Figure 5-11 shows this fault current development. Because of this effect, an additional 16% coordination margin should be considered between primary and secondary transformer protective devices.

As illustrated by Figure 5-10, to coordinate a LVPCB with devices on its load side, the circuit breaker's time-current curve must not cross the time-current curve of the downstream (load-side) device up to the maximum available short-circuit current. Should the time-current curves cross, a potential lack of selective coordination exists.

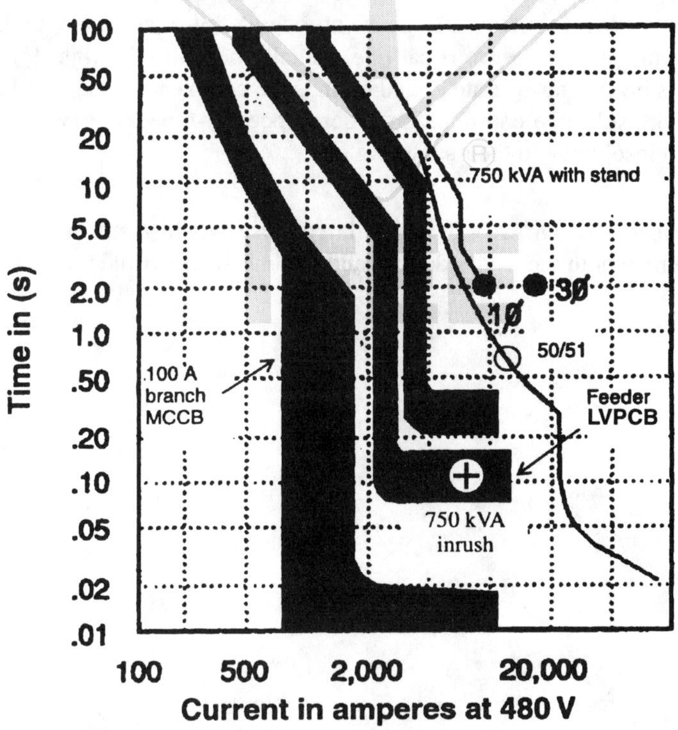

Figure 5-10—Low-voltage circuit breakers and primary overcurrent relay

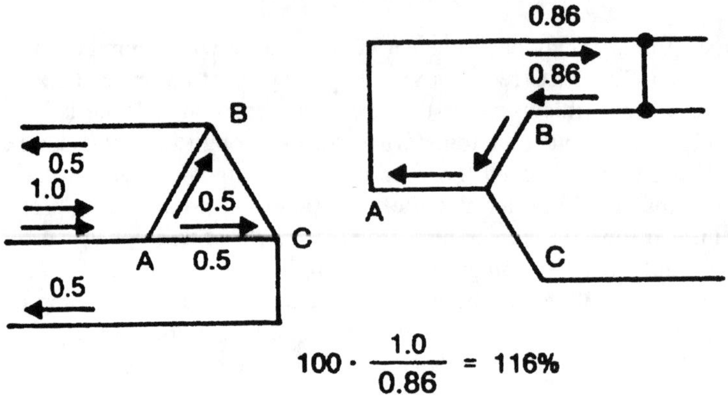

Figure 5-11—Currents in delta-wye transformer for secondary phase-to-phase fault

5.5.2 Fuse and circuit breaker

Figure 5-12a shows apparent selective coordination between a thermal-magnetic circuit breaker and a load-side current limiting fuse, represented in the figure by its total clearing curve. To ensure coordination for critical applications, it is also necessary to examine the sub-half cycle region (times less than 0.01 s). The no-operate lower current boundary of the magnetic element, while shown as a vertical line, actually bends to the right for times less than 0.01–0.02 s. This inverse no-operate boundary must be compared to the fuse let-through characteristic to ensure selective coordination. Microprocessor-type trip devices may have to be examined more closely due to their sampling rate.

Figure 5-12b shows that an MCCB with a larger fuse would result in a nonselective area above 3000 A where both the load-side fuse and the line-side thermal magnetic MCCB would open.

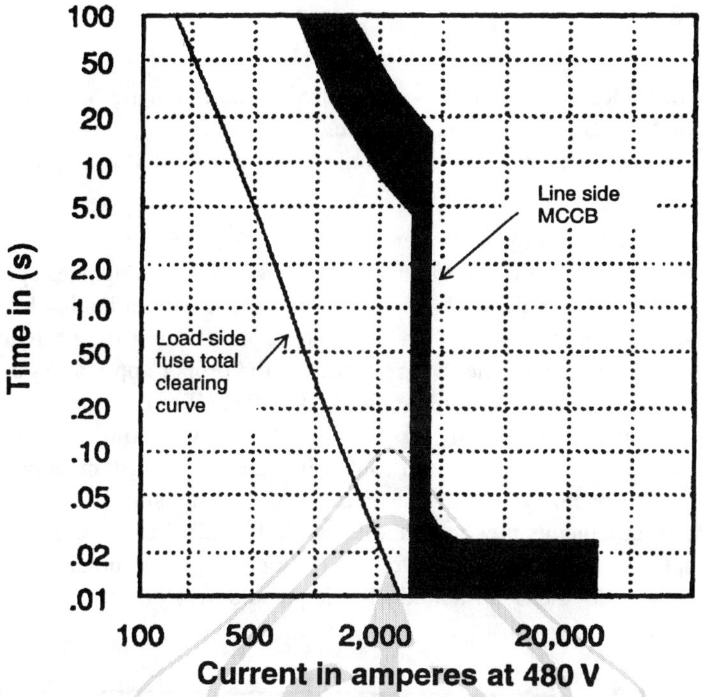

Figure 5-12a—MCCB and load-side fuse

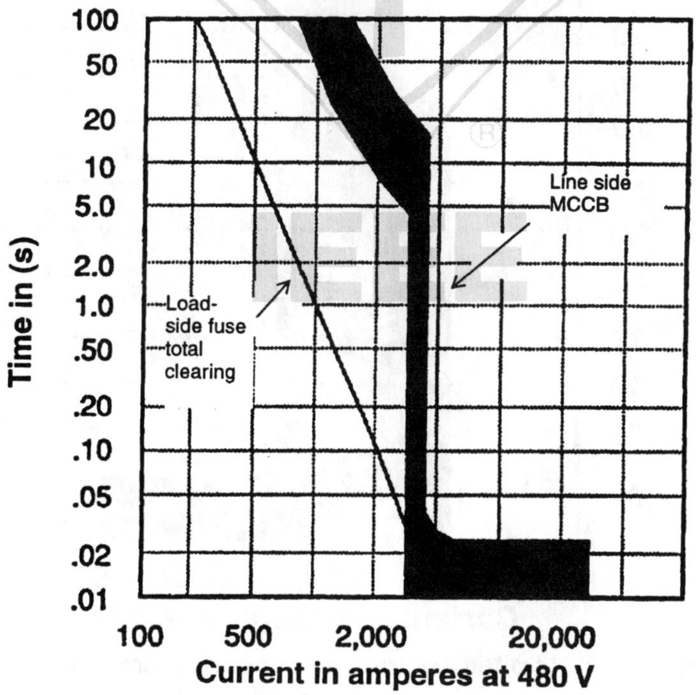

Figure 5-12b—Miscoordination of circuit breaker and load-side fuse

5.5.3 Series MCCBs

Figure 5-13 shows selectivity between a line-side thermal magnetic MCCB and a load-side thermal magnetic MCCB up to 3000 A available. For this situation, overcurrents up to 3000 A will open just the load-side circuit breaker, while faults above 3000 A may open both.

MCCBs in series, recognized by UL as series-connected devices, have limited selective coordination, essentially as illustrated by Figure 5-13. UL 489-1991 recognizes specific combinations of two MCCBs (and of line-side current-limiting fuses and a load-side MCCB) for an available short-circuit current higher than the interrupting rating of the load-side MCCB. Refer to Chapter 3 of this recommended practice for the proper application of series combination devices. For a recognized combination of two MCCBs, the line-side circuit breaker must have an instantaneous trip setting less than the interrupting rating of the load-side circuit breaker. Selective coordination is limited to currents below the instantaneous pickup of the line-side circuit breaker. For any fault downstream of the load-side MCCB having a current greater than the instantaneous pickup of the line-side MCCB, both circuit breakers trip, and power is interrupted to unfaulted circuits fed by the line-side circuit breaker. In a fully coordinated system, power to these unfaulted circuits would not be interrupted, except for a momentary voltage dip.

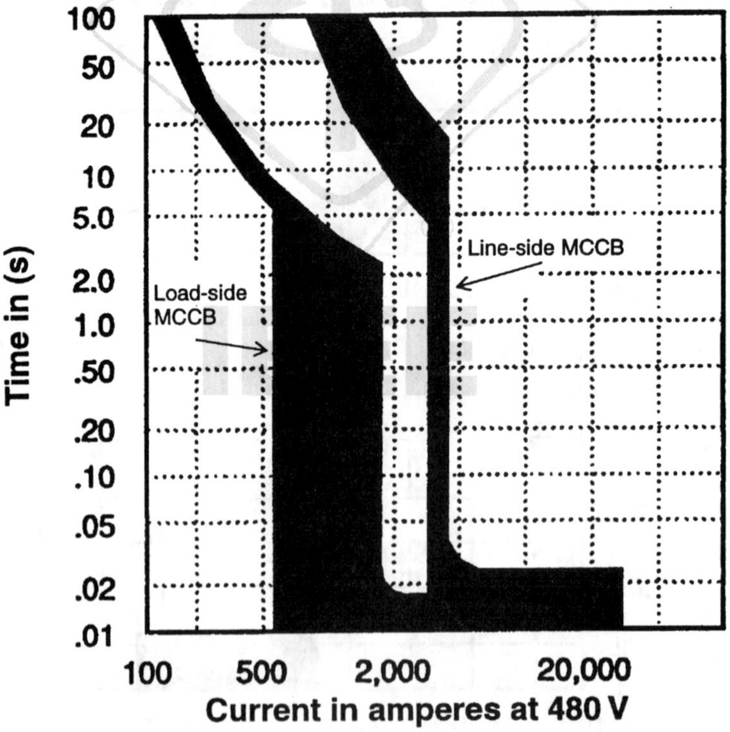

Figure 5-13—Limited coordination of series circuit breakers

5.5.4 Short-time delay

If the MCCB of Figure 5-12b is replaced by a similar device with electronic trips including a short-time delay unit, selectivity is improved as in Figure 5-14. The addition of a short-time delay provides selective coordination up to the circuit breaker's instantaneous pickup, or override.

Figure 5-15 depicts the increased selectivity afforded by the use of short time delay on a line-side MCCB with a load-side, thermal-magnetic trip MCCB. Coordination is achieved up to 7500 A where the instantaneous override takes over on the MCCB. Without short-time delay, coordination was achievable up to only 3000 A available (see Figure 5-13).

All MCCBs and most ICCBs with a short-time delay have an instantaneous override. If selective coordination is required for fault currents above the instantaneous override level, a power circuit breaker defined as an LVPCB with a short-time delay option, as shown in Figure 5-16, should be considered.

Referring back to Figure 5-8 shows how total selectivity between the line-side power circuit breaker with short-time delay and the load-side thermal-magnetic trip MCCB can be achieved. All power system components between the power circuit breaker with the short-time trip and the load-side MCCB device must be capable of withstanding the flow of the maximum fault current for several cycles of short-time delay before tripping.

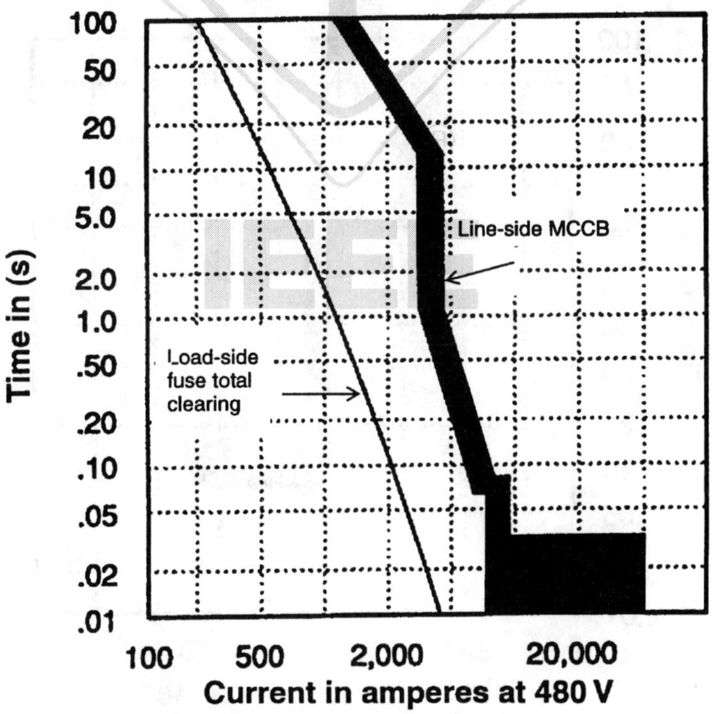

Figure 5-14—MCCB with short-time delay coordinating with load-side fuse

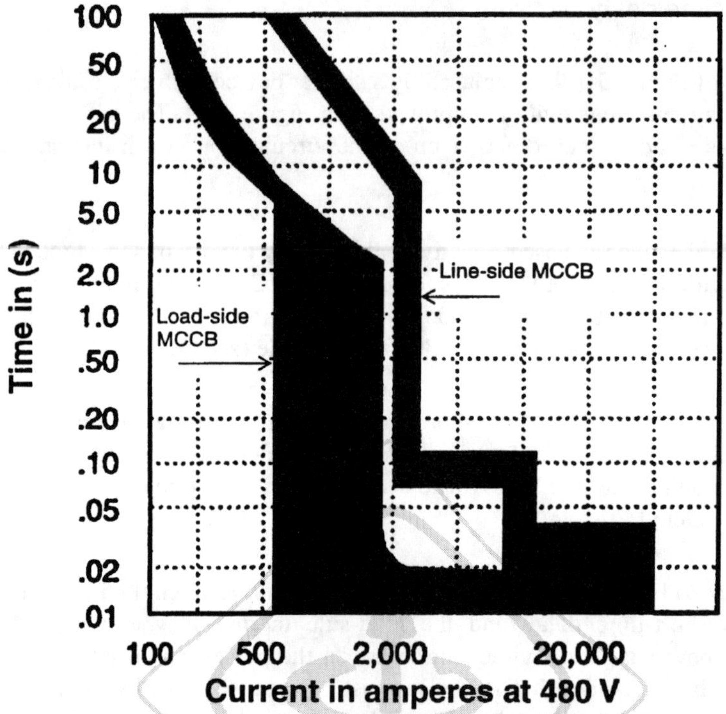

Figure 5-15—MCCB with short-time delay and load-side MCCB

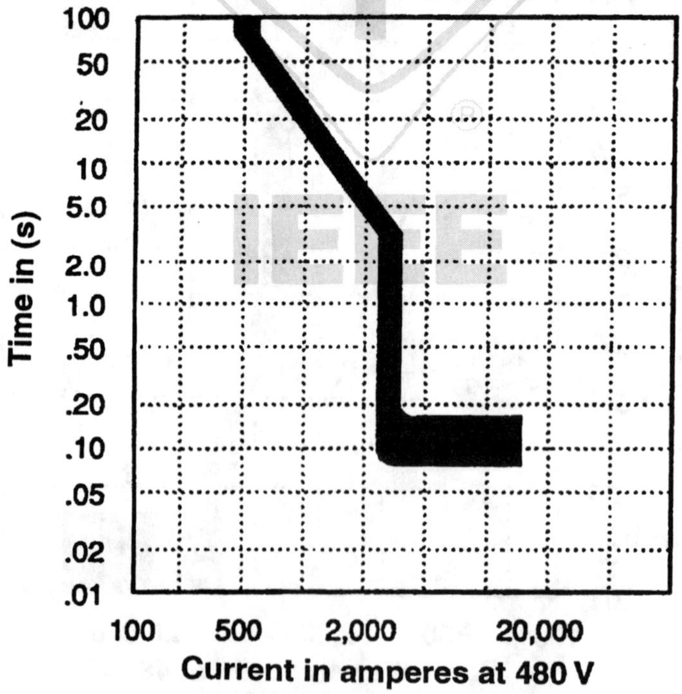

Figure 5-16—LVPCB with short-time delay

5.5.5 Current-limiting circuit breakers

Figure 5-17 shows a typical time-current curve for a current-limiting MCCB. Figure 5-18 shows the additional information (sub-half-cycle unlatching time) required to draw the full time-current curve so that a coordination study can be performed. As with standard circuit breakers, current-limiting circuit breakers will coordinate with other line-side and load-side devices as long as the time-current curves do not cross. If the curves do cross, coordination is achieved up to the point where the curves intersect. When coordinating a current- limiting circuit breaker or those with dynamic impedance, the manufacturer should be consulted to determine what effect the circuit breaker will have on the fault current that flows through other service devices. Dynamic impedance is a term used to describe devices that quickly establish an impedance in the faulted circuit to limit current, and therefore limit component damage.

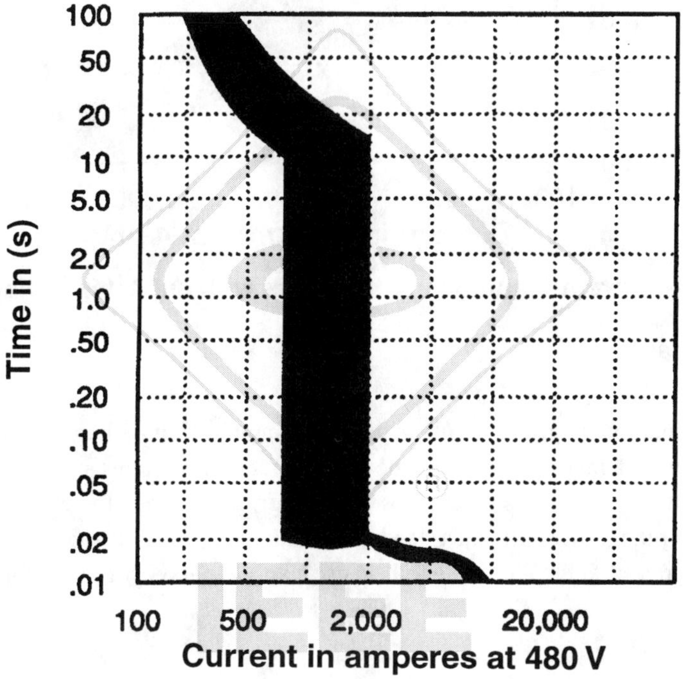

Figure 5-17—Current-limiting circuit breaker

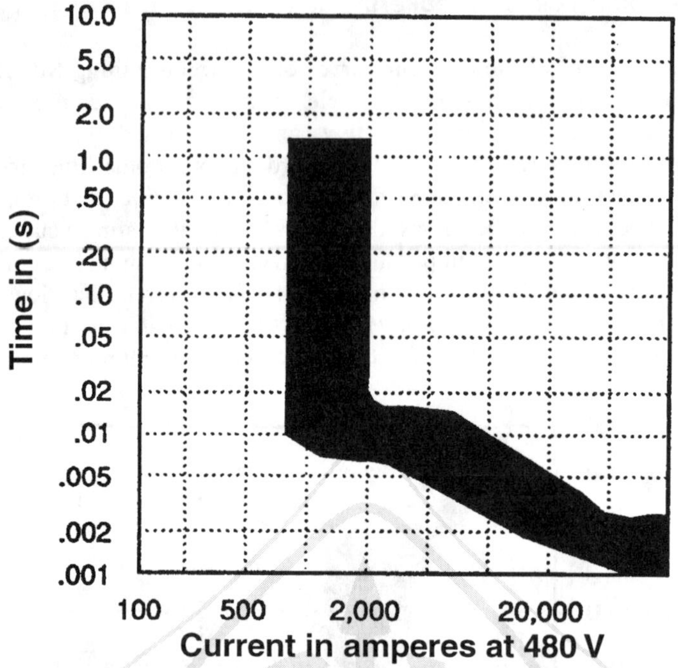

Figure 5-18—Current-limiting circuit breaker with unlatching characteristic

5.5.6 Ground fault

Selective coordination of ground-fault trip devices among low-voltage circuit breakers is readily accomplished in most cases as the trip characteristics consist of several bands of nearly constant time (see Figure 5-19). It is also important to check the characteristics of the ground-fault tripping device of each feeder with the largest load-side phase overcurrent device without ground fault. This is done because a single phase-to-ground fault will involve those two devices.

When a large fuse, or motor starter without a ground-fault trip device, is used on the load (as shown in Figure 5-19), a judgment may be required to use an I^2t pickup on the ground-fault trip (of course checking the impact on line-side devices), raising the ground pickup setting, raising the ground delay setting, some combination of these, or accepting some miscoordination. There are a few systems that require special attention—double-ended four-wire ground, zone interlocking, and arcing ground fault memory, to name a few. The manufacturer's application information should be consulted. Another area that warrants additional consideration in systems with high ground-fault currents is coordinating the ground-fault tripping characteristic of a load-side device with both the ground-fault and the phase overcurrent trip characteristics of the next line-side device.

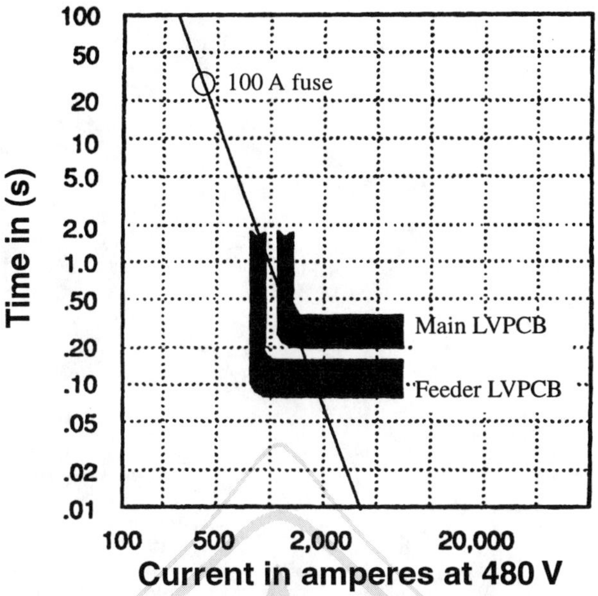

Figure 5-19—Main and feeder circuit-breaker ground-fault coordination with load-side fuse

5.6 References

This chapter shall be used in conjunction with the following publications:

ANSI C37.17-1979 (Reaff 1988), American National Standard for Trip Devices for AC and General Purpose DC Low-Voltage Power Circuit Breakers.[2]

IEEE Std 141-1993, IEEE Recommended Practice for Electric Power Distribution for Industrial Plants (*IEEE Red Book*) (ANSI).[3]

IEEE Std 241-1990, IEEE Recommended Practice for Electric Power Systems in Commercial Buildings (*IEEE Gray Book*) (ANSI).

IEEE Std 242-1986 (Reaff 1991), IEEE Recommended Practice for Protection and Coordination of Industrial and Commercial Power Systems (*IEEE Buff Book*) (ANSI).

NFPA 70-1996, National Electrical Code® (NEC®).[4]

UL 489-1991, Molded-Case Circuit Breakers and Circuit-Breaker Enclosures (DoD).[5]

---

[2]ANSI publications are available from the Sales Department, American National Standards Institute, 11 West 42nd Street, 13th Floor, New York, NY 10036, USA.

[3]IEEE publications are available from the Institute of Electrical and Electronics Engineers, 445 Hoes Lane, P.O. Box 1331, Piscataway, NJ 08855-1331, USA.

[4]The NEC is available from Publications Sales, National Fire Protection Association, 1 Batterymarch Park, P.O. Box 9101, Quincy, MA 02269-9101, USA. It is also available from the Institute of Electrical and Electronics Engineers, 445 Hoes Lane, P.O. Box 1331, Piscataway, NJ 08855-1331, USA.

[5]UL publications are available from Underwriters Laboratories Inc.,333 Pfingsten Road, Northbrook, IS 60062-2069, USA.

IEEE

Chapter 6
Special-purpose circuit breakers

6.1 Introduction

This chapter covers five types of special-purpose circuit breakers. These are circuit breakers other than conventional molded-case circuit breakers (MCCBs) or low-voltage power circuit breakers (LVPCBs). The chapter acknowledges that there exist a wide variety of special circuit-breaker products, and that the five selected here do not comprise the complete selection of special-purpose circuit breakers and their derivatives.

In this chapter, the following special-purpose circuit breakers and their derivatives will be discussed:

— Instantaneous-trip circuit breakers
— Mine-duty circuit breakers
— Current-limiting circuit breakers
— Molded-case switches
— Integrally fused circuit breakers

6.2 Instantaneous-trip circuit breakers

Instantaneous-trip circuit breakers (motor circuit protectors) provide adjustable short-circuit protection, but no overload protection. Since external overload protection must be used with these breakers, they cannot be used for branch circuit protection. These breakers are primarily used as components in motor circuits in combination with motor starters to provide the short-circuit protection function. They may also serve as the motor disconnecting means. The most typical applications are in motor control centers or individual combination motor starters. They are also used in welding equipment for short-circuit protection only. Figure 6-1 shows typical instantaneous-trip circuit breakers.

6.2.1 Ratings

Instantaneous-trip circuit breakers have a maximum continuous-current-carrying capacity. Prolonged continuous currents exceeding this rating may cause damage to the circuit breaker due to overheating.

Instantaneous-trip circuit breakers do not carry an interrupting rating by themselves under industry standards. Instantaneous-trip circuit breakers are most often applied in conjunction with motor starters. Combination starters are short-circuit tested with the starter and instantaneous-trip circuit breaker installed. The short-circuit rating that is applied is marked on the combination starter as a result of this test.

NOTE—Per UL 508-1993[1] test requirements, after a short circuit has occurred the circuit breaker and wires must be capable of continued service. However, the overload relay, contactor, or both, may require repair or replacement.

[1]Information on references can be found in 6.7.

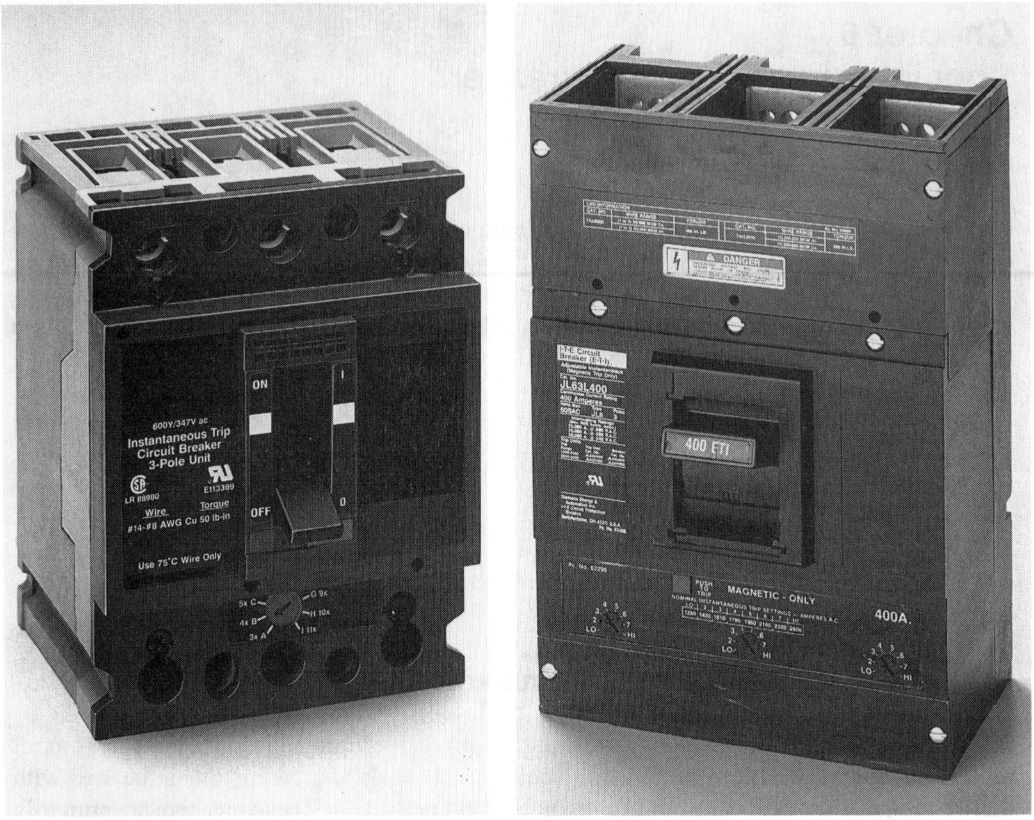

Sources: Square D Company (left photo), Siemens Energy & Automation, Inc. (right photo).

Figure 6-1—Examples of instantaneous-trip circuit breakers

UL component recognized breakers are short-circuit tested by themselves at limited available (standard) levels as a basic requirement even though they do not carry an interrupting rating.

6.2.2 Current-limiting attachments

Several manufacturers have add-on current-limiting attachments that, if added to the load-side of the instantaneous-trip circuit breaker, will significantly raise the short-circuit withstand current rating of the combination starter. Current-limiting attachments contain specially designed fuses that open and limit the current that flows under high-level fault conditions. Low-level fault currents will be interrupted by the instantaneous-trip circuit breaker without opening the current-limiting attachment fuses. High-level fault currents are interrupted by both the current-limiting module and the circuit breaker. The current-limiting attachment must be replaced after interrupting.

6.2.3 Code considerations

The National Electrical Code® (NEC®) (NFPA 70-1996), Section 430-52 states that

> "An instantaneous-trip circuit breaker shall be used only if adjustable and if part of a listed combination controller having coordinated motor overload and short-circuit and ground-fault protection in each conductor, and the setting is adjusted to no more than the value specified in Table 430-152. A motor short-circuit protector shall be permitted in lieu of devices listed in Table 430-152 if the motor short-circuit protector is part of a listed combination controller having coordinated motor overload protection and short-circuit and ground-fault protection in each conductor and it will open the circuit at currents exceeding 1300 percent of full-load motor current.
>
> (FPN): For the purpose of this article, instantaneous-trip circuit breakers may include a damping means to accommodate a transient motor inrush current without nuisance tripping of the circuit breaker.
>
> Exception No. 1: Where the setting specified in Table 430-152 is not sufficient for the starting current of the motor, the setting of an instantaneous-trip circuit breaker shall be permitted to be increased but shall in no case exceed 1300 percent of the motor full-load current for other than Design E motors and no more than 1700 percent of full-load motor current for Design E motors. Trip settings above 800 percent for other than Design E motors and above 1100 percent for Design E motors shall be permitted where the need has been demonstrated by engineering evaluation. In such cases, it shall not be necessary to first apply an instantaneous-trip circuit breaker at 800 percent or 1100 percent.
>
> Exception No. 2: Where the motor full-load current is 8 amperes or less, the setting of the instantaneous-trip circuit breaker with a continuous current rating of 15 amperes or less in a listed combination motor controller that provides coordinated motor branch-circuit overload and short-circuit and ground-fault protection shall be permitted to be increased to the value marked on the controller."

NEC Table 430-152 identifies the maximum rating or setting of the instantaneous-trip circuit breaker as 800% for motors other than Design E, and as 1100% for Design E motors other than the above exceptions.

Design E motors are energy-efficient motors. It is important to note that there are Design B energy-efficient motors that are not permitted to have trip settings higher than 1300% of full-load current and some constant horsepower multispeed motors that may cause instantaneous-trip circuit breakers to trip even at 1300% of the full-load current. The use of instantaneous-trip circuit breakers in these applications is not recommended.

6.2.4 Setting of instantaneous-trip circuit breakers

When used in a combination starter, the overload relays provide time delay for starting as well as overcurrent protection up to the locked rotor motor current range. By setting the instantaneous-trip circuit breaker trip level just above the motor starting current, maximum protection can be achieved without nuisance tripping on startup. Most manufacturers publish recommended continuous-current ratings and magnetic-trip ranges for commonly available motors.

When information is not otherwise available and the motor *is* available, starting current can be estimated by taking the motor code from its nameplate and using the code letter to estimate locked rotor current from NEC Table 430-7(b). Starting current in rms amperes will be approximately two times locked rotor current.

When information is not otherwise available and the motor is *not* available, locked rotor current can be estimated by using NEC Table 430-151 A or B. Starting current in rms amperes can be estimated to be two times locked rotor current.

6.2.5 Energy-efficient motors

Energy efficient Design B motors may exhibit starting currents higher than 1300% of their full-load current. At the time of this writing, the NEC does not address this situation. The recommended solution is to use a thermal-magnetic circuit breaker that can be applied within NEC Table 430-152 and that has an Instantaneous-trip level sufficient for the motor starting current.

6.3 Mine-duty circuit breakers

Mine-duty circuit breakers are specifically designed for mining duty applications and permit the user to comply with mandatory mine-duty standards. The normal operation of self-propelled mining equipment subjects its trailing cable to extreme and frequent flexing, twisting, and crushing. As a result, electrical faults in trailing cables occur much more frequently than wiring in normal, stationary installations. Additionally, the presence of loose coal dust and other combustible materials makes the occurrence of such faults very hazardous. For these reasons, adequate trailing cable protection is extremely important.

Mine-duty circuit breakers typically have adjustable Instantaneous-trip settings with tighter tolerances, dust shields and gaskets, non-moisture-absorbing materials, heavy-duty operating mechanisms, heavy-duty undervoltage releases with an external push button, and corrosion-resistant nameplates. Mine-duty circuit breakers are available with voltage ratings up to 1000 V ac and 300 V dc.

6.3.1 Magnetic trip setting

CFR, Title 30, Section 75.601, Part 75 indicates that

> "Circuit breakers providing short circuit protection for trailing cables shall be set so as to not exceed the maximum allowable instantaneous settings specified in this section; however, higher settings may be permitted by an authorized representative of the Secretary when he has determined that special applications are justified."

Mine-duty circuit breakers offer adjustable magnetic-trip points or at least several settings that comply with the required maximum allowable trip settings in Table 6-1.

Table 6-1—Maximum allowable circuit-breaker instantaneous setting

Conductor size (AWG or kcmil)	Setting (A)
14	50
12	75
10	150
8	200
6	300
4	500
3	600
2	800
1	1000
1/0	1250
2/0	1500
3/0	2000
4/0	2500
250	2500
300	2500
350	2500
400	2500
450	2500
500	2500

Source: CFR, Title 30, Chapter 1, Section 75.601-1.

6.3.2 Testing requirements

CFR, Title 30, Section 75.900-2, Part 75 states that

> "Circuit breakers protecting low- and medium-voltage alternating current circuits serving three phase alternating current equipment and their auxiliary devices shall be tested and examined at least once each month by a person qualified as provided in 75.153. In performing such tests, actuating any of the circuit breaker auxiliaries or control circuits in any manner which causes the circuit breaker to open, shall be considered a proper test. All components of the circuit breaker and its auxiliary devices shall be visually examined and such repairs or adjustments as are indicated by such tests and examinations shall be carried out immediately."

6.3.3 Circuit breaker location

MCCBs used to protect underground circuits are required to be located in areas that are accessible for inspection and testing and have a safe roof.

6.3.4 Additional application requirements

Mine-duty circuit breakers feeding three-phase ac circuits are required to be equipped with devices to provide protection against undervoltage, grounded phase, short circuits, and overcurrent.

Frequently, mine-duty circuit breakers are specified to be equipped with an undervoltage release. The undervoltage release can serve the following three different purposes:

a) To trip the circuit breaker during an undervoltage or power outage condition
b) To provide a tripping mechanism that can be used with ground-fault or interlock circuits
c) To provide an emergency tripping device that can be activated by a remote switch or push-button

6.4 Current-limiting circuit breakers

UL 489-1991 defines a current-limiting circuit breaker as follows:

> "A circuit breaker that does not employ a fusible element and that when operating within its current limiting range, limits the let-through I^2t to a value less than the I^2t of a 1/2 cycle wave of the symmetrical prospective current."

6.4.1 Description

Current-limiting circuit breakers are available from 15 A to 1200 A, rated up to 600 V, and have interrupting ratings up to 200 000 rms symmetrical A. These circuit breakers completely clear the faulted circuit within the first half cycle.

Current-limiting circuit breakers are basically conventional thermal-magnetic or electronic-trip circuit breakers, designed so that high-speed contact separation is achieved under high-level fault conditions. This high-speed contact separation effectively limits potentially high-level fault currents and is achieved by closely spaced parallel contact arms carrying current in opposite directions. One form of a current-limiting circuit breaker with high-speed contacts is illustrated in Figure 6-2.

Current-limiting circuit breakers can be reset and service can be restored in the same manner as conventional circuit breakers even after clearing maximum-level fault currents. Of course, whenever a fault has occurred, it is important to remove the fault and its cause before reenergizing. If indications are that the fault was a significant short-circuit, the circuit breaker should be examined before reenergizing. If there are cracks in the circuit-breaker housing, if operation is difficult, or if there is severe discoloration, the circuit breaker should be replaced before reenergizing. NEMA AB 4-1996 contains more detailed information on testing and inspecting circuit breakers that have been in service.

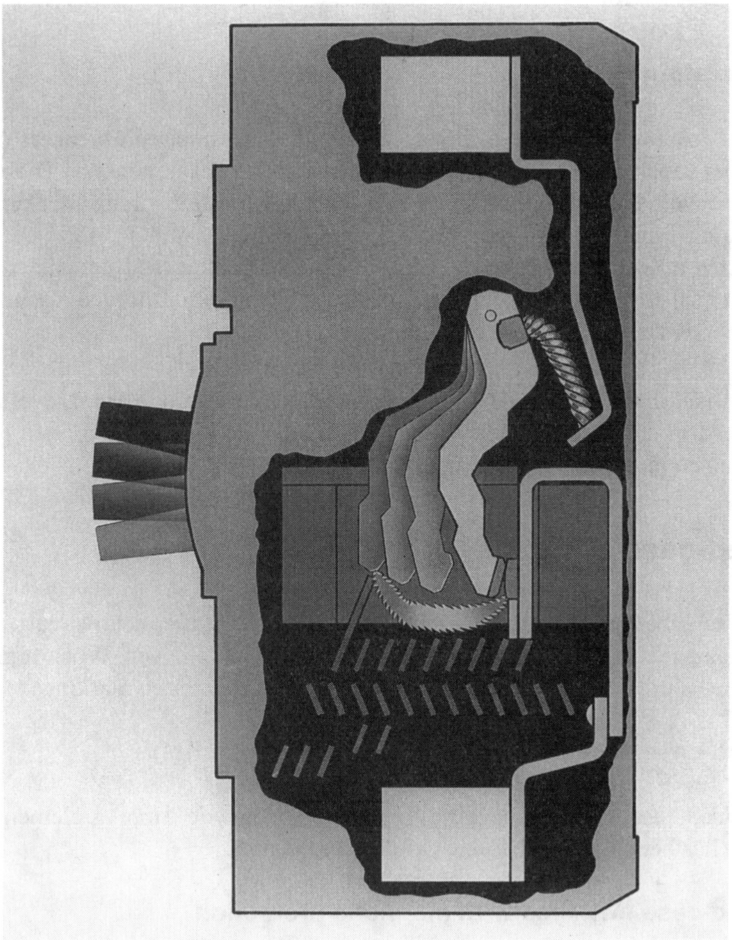

Source: Square D Company.

Figure 6-2—Current-limiting circuit breaker

6.4.2 Applications

Current-limiting circuit breakers not only provide high interrupting ratings, but also limit let-through current and energy to load-side devices and conductors.

The current-limiting action is sufficient to reduce I_p and I^2t let-through to values that lesser-rated downstream circuit breakers can interrupt. A common application of this is using the current-limiting circuit breaker as either an integral or remote main breaker for lighting panelboards on systems with high available fault currents. This is called a series rating. It is important to note that manufacturers must test to demonstrate that the downstream device is protected in order to achieve a UL Recognized series rating. There is presently no accurate method of calculating this protection.

6.4.3 Series ratings

UL Recognized series ratings are available utilizing upstream circuit breakers or fuses with current-limiting capability in series with lesser-rated downstream breakers. These recognized combinations have been tested to verify that the current-limiting circuit breaker will protect the downstream device under the test conditions. Individual manufacturers' UL Recognized series-connected ratings are published and may be found in the UL Recognized Component Directory [B1][2]. It is important to note that series ratings are not limited to current-limiting circuit breakers. Refer to Chapter 3 for additional information.

It should be noted that selectivity will not be provided above any current level at which the circuit breaker trip characteristic curves overlap. Where continuity of service is desired, or required, series-connected ratings are not recommended.

6.5 Molded-case switches

Molded-case switches are essentially circuit breakers with the thermal overcurrent protection removed. The magnetic short-circuit protection may also be removed. When high-level magnetic protection is provided, that fact will be indicated on the switch markings.

6.5.1 Molded-case switches without magnetic protection

Standard molded-case switches have no thermal- or magnetic-tripping elements. The term *nonautomatic* has been used with these switches in the past.

6.5.2 Molded-case switches with magnetic protection

Molded-case switches with magnetic-trip elements do not provide overcurrent protection. However, they do include a preset nonadjustable magnetic-trip element that serves to protect the switch against the damaging effects of high-level fault currents. For most applications, the standard (nonautomatic) and automatic switches can be used interchangeably. In considering the difference, the automatic switch will provide a second level of isolation while the standard switch does not have to be reclosed and may be desirable in some logic systems.

6.5.3 Ratings

Molded-case switches have a maximum continuous-current rating which, if exceeded for long periods of time, may cause damage to the switch due to overheating. Neither standard nor automatic molded-case switches have interrupting ratings, as they are intended only to be a disconnecting switch, not a protective device. Molded-case switches must be protected by an overcurrent protective device of an equivalent or lower current rating. Even when protected by an overcurrent protective device, molded-case switches should not be applied on systems capable of delivering fault current in excess of their withstand rating. Molded-case switches may carry a withstand rating from 5000 A to 200 000 A and are labeled for use in

[2]The numbers in brackets correspond to those of the bibliography in 6.8.

series with a specific overcurrent protective device or devices. The short-circuit rating test is conducted with the switch in series with the overcurrent device or devices specified.

6.5.4 Applications

Molded-case switches provide a simple and compact disconnecting means. Additionally, they meet the requirements of NEC Section 430-109:

> "The disconnecting means shall be one of the following types: a motor circuit switch rated in horsepower, a circuit breaker, or a molded case switch, and shall be a listed device."

Molded-case switches are capable of making and breaking load currents up to six times their marked rating. As a result, these switches can be applied where horsepower ratings are required.

6.6 Integrally fused circuit breakers

Integrally fused circuit breakers are available in both molded-case and power circuit breaker constructions. They provide high interrupting capability through the use of specially designed current-limiting fuses, termed *limiters* in this application to distinguish them from commercially available Class R, J, or L fuses. Such fuses are assembled into the housing of the circuit breaker or, in the case of high-ampere power circuit breaker frames, in a separate fuse truck. The limiters in these devices are designed to open, and need replacement, only after a high-level fault. The circuit-breaker portion is interlocked so that when any limiter opens, the circuit breaker will automatically trip, opening all poles of the circuit breaker and eliminating the possibility of single phasing caused by the opening of one of the limiters. Additionally, many circuit breakers are equipped with a mechanical interlock that prohibits the circuit breaker from closing with a missing limiter.

In the MCCB construction, the limiters are generally located within an added housing and are separated from the sealed trip unit of the circuit breaker for easy access. In power circuit breaker construction, the limiters are mounted on the rear of the circuit-breaker frame or in a separately mounted fuse truck. Both mounting methods provide easy access to the limiters.

6.6.1 Applying fused circuit breakers

For ideal coordination within the fused circuit breaker, fuses should be selected so that overcurrents and low-magnitude faults are cleared by thermal or long-time/short-time tripping action; intermediate-level short circuits are cleared by short-time delay, magnetic, or instantaneous tripping action; high-level short circuits are cleared by the current-limiting fuse and instantaneous tripping action. This selection is usually made by the manufacturer, although optional fuse selections are sometimes available that may affect this coordination.

The short-time delay characteristic is beneficial if coordination in the intermediate current range is desired. There is a lack of coordination with the load-side protective devices when currents are in the current-limiting range of the fuse.

When applied on high short-circuit current capacity systems, the effects of the let-through characteristics of the fused circuit breakers on downstream equipment must be considered. The presence of the current-limiting fuse as part of the fused circuit breaker does not necessarily imply that the downstream equipment can adequately withstand these effects. It is important to note that manufacturers must test to ensure that in all combinations the downstream device is protected. There is no accurate method of calculating this protection.

Fused circuit breakers may be used where very high fault currents are available. In addition, some fused MCCBs provide series short-circuit ratings with other MCCBs. These series ratings are higher than the load-side circuit-breaker rating.

It should be noted that fused circuit breakers do not have any current-limiting effect until the current associated with the fault exceeds the threshold current of the limiter. Using fused breakers to protect standard breakers downstream is sound design, provided the manufacturer's coordination data is applied carefully. It needs to be understood that reduced size limiters may be necessary, but the use of these limiters could result in less than ideal coordination within the fused circuit breaker. This may cause frequent blowing of the limiters. In no case should combinations of trip devices and fuses that are not approved by the manufacturer be installed. Where fuses of different manufacture are being considered for the same system, the characteristics and peak let-through current of a given fuse rating may vary substantially between manufacturers.

It is important to note that electrical distributors who stock ordinary current-limiting fuses rarely stock these limiters. To distinguish these fuses from more common current limiting fuses, they are most typically termed *limiters* by standards such as UL 489-1991 and by many distributors. It is wise to consider keeping at least one set of spare limiters of each size used.

6.7 References

This chapter shall be used in conjunction with the following publications:

Code of Federal Regulations (CFR), Title 30—Mineral Resources, Chapter 1—Mine Safety and Health Administration. (USA).[3]

NEMA AB 3-1996, Molded Case Circuit Breakers and Their Application.[4]

NEMA AB 4-1996, Guidelines for Inspection and Preventive Maintenance of Molded Case Circuit Breakers Used in Commercial and Industrial Applications.

NFPA 70-1996, National Electrical Code®(NEC®).[5]

UL 489-1991, Molded-Case Circuit Breakers and Circuit-Breaker Enclosures (DoD).[6]

UL 508-1993, Industrial Control Equipment (DoD).

UL 1087-1993, Molded-Case Switches.

6.8 Bibliography

Additional information may be found in the following source:

[B1] UL Recognized Component Directory, 1996.

[3]The Code of Federal Regulations is available from the Superintendent of Documents, U. S. Government Printing Office, Washington, DC 20037, USA.

[4]NEMA publications can be obtained from the National Electrical Manufacturers Association, 1300 N. 17th Street, Suite 1847, Rosslyn, VA 22209, USA.

[5]The NEC is available from Publications Sales, National Fire Protection Association, 1 Batterymarch Park, P.O. Box 9101, Quincy, MA 02269-9101, USA. It is also available from the Institute of Electrical and Electronics Engineers, 445 Hoes Lane, P.O. Box 1331, Piscataway, NJ 08855-1331, USA.

[6]UL standards are available from Global Engineering, 1990 M Street NW, Suite 400, Washington, DC, 20036, USA.

IEEE

Chapter 7
Acceptance and maintenance requirements

7.1 Scope

This chapter provides guidelines and instructions for developing an acceptance and maintenance testing program for low-voltage circuit breakers. Such items as maintenance planning, establishment of baseline data, training considerations, safety considerations, testing considerations, acceptance criteria, and documentation requirements are discussed.

7.2 Maintenance program

The time to begin planning for maintenance of circuit breakers, or any kind of equipment, is when the equipment is being installed. It is good practice for the organization that will be responsible for maintenance to be present, if possible, when the circuit breakers and associated equipment are initially installed and checked out. This initial familiarity will help maintenance personnel feel more comfortable with the equipment and provide information that will save time and questions when maintenance or troubleshooting becomes necessary.

If an organization plans on maintaining the LVCBs themselves, they should obtain the proper training for their personnel. Such training is often available from the manufacturer or other service organizations that are dedicated to maintenance activities.

The first order of business of any viable LVCB maintenance program is preparing a detailed procedure based on what has been learned from initial installation, test data, manufacturer's information, and training. The procedure should provide a step-by-step guide to performing the desired level of maintenance. The procedure should provide identifiable guidelines and acceptance criteria such that maintenance personnel will know what is and is not acceptable.

Different documents exist that will aid in developing this detailed procedure, including the following:

a) Single line diagrams
b) Elementary (control) diagrams
c) Relay setting sheets
d) Circuit breaker manufacturer trip curves
e) Technical requirement for specifications
f) Vendor wiring/connection diagram (as applicable)
g) Vendor instruction manual (as applicable)
h) Personnel qualifications, certifications, and training

A maintenance schedule for LVCBs should be established. The schedule might be based on the manufacturer's recommendations and may be adjusted based on previous experience

(Appendix H4 of NFPA 70B-1994[1] provides good maintenance guidelines for all low-voltage equipment). The schedule will also be dependent on production demands, number of LVCB operations, and the environment in which the breaker is located. Of these three, the environment in which an LVCB is located is most important.

Environmental conditions encountered in the field must be evaluated to determine the optimum frequency for the performance of maintenance. Some pieces of equipment exhibit characteristics that require maintenance procedures to be performed on a more frequent basis to ensure proper operation regardless of the environment in which they are placed. When a circuit breaker is placed in a clean, well conditioned environment it may sit idle for a significant period of time with little or no adverse effects and operate properly when required.

When a circuit breaker is in a dusty, dirty, or corrosive atmosphere it might either become inoperative or it may malfunction if it should attempt to interrupt a fault that is well within its capacity. The frequency of preventive maintenance, inspection, and cleaning must be high in order to ensure the integrity of operation.

Once the appropriate technical procedures and maintenance schedules have been prepared, safety aspects of the job must also be considered before the actual work is performed. Procedures and schedules may have to be modified as a result.

The safety of personnel performing maintenance inspection, testing, and repair should be a prime requirement of any maintenance program. Persons performing periodic maintenance on LVCBs should be familiar with the electrical system(s) in which the breakers are used. They should know where and how the breaker they are working on is isolated from energized portions of the circuit. Testing should be performed before proceeding with any maintenance work to ensure that all sources of electrical potential have been removed. This type of work should be performed by trained personnel only.

Low-voltage power circuit breakers (LVPCBs) shall be totally disconnected from the switchgear bus connections before performing tests or checks on the actual breaker. If work on the racking mechanism or any components within the breaker cubicle is to be performed, isolation of the line-side stabs is necessary. Some circuit configurations exist such that the load-side terminals are energized from another source. This type of condition is known as a back-feed. In such a case, isolation of the load-side stabs would also be required before working within the cubicle. Where possible, the preferred method of stab isolation is to de-energize the circuit remote from the cubicle location. If de-energization is not possible, the stabs may be covered. Most manufacturers provide covers or isolators for the stabs. However, extreme caution must be exercised when covering energized stabs.

Even after LVPCBs are withdrawn from the cubicle to test and observe the operation, there may still be a hazard due to the stored energy mechanisms. Awareness must be maintained throughout the duration of the testing so that hands and fingers are kept clear of pinch points on LVPCBs.

[1]Information on references can be found in 7.8.

The only method of total isolation of molded-case circuit breakers (MCCBs) is de-energization of the source and load. Removal of an MCCB from a panel must not be done if the panel remains energized, unless the breaker is of a plug-on or draw-out type construction.

7.3 Maintenance of MCCBs

Field maintenance of MCCBs is normally limited to a visual inspection, cleaning, and tightening of connections. Any MCCB found to have a cracked case when inspected should be replaced because of possible interrupting capability loss.

One other observable phenomenon that requires attention is tracking. Tracking is an electrical discharge phenomenon indicated by a leakage path that is directionally erratic (similar to the pattern of a lightning stroke). This phenomenon forms from electrical stress over a long period of time, especially in unclean environments, and will eventually lead to a flashover.

MCCBs are designed and manufactured to require no internal maintenance throughout their lifetime. MCCBs are factory calibrated and sealed, and the breaker should not be tampered with. However, some MCCBs have a removable cover that is not intended only for installing trip units. This does not mean that maintenance can be performed on the MCCB with the cover off. The manufacturer did not intend this to be the case.

The following items should be checked on an MCCB during periodic maintenance schedule:

a) *Breaker overheating*. Check the breaker for overheating when it is operating under normal load at a normal operating temperature. Check the face of the breaker using the palm of the hand. If hand contact cannot be maintained for 3 s, the breaker should be removed from service and checked further to determine the cause of the overheating. Thermal imaging equipment is also available to determine overheating conditions.

b) *Connection check*. Visually check the breaker and bus/cable connections for evidence of overheating. Overheated copper connections can usually be cleaned and dressed satisfactorily for reuse. Overheated aluminum connectors must be replaced. All connections must be tightened to proper torque levels specified after cleaning or replacement.

c) *Mechanical operation*. All MCCBs should be exercised (opened and closed) several times to ensure proper mechanical operation. Exercising an aged breaker is particularly important, considering many breakers are never called upon to operate by an overcurrent condition. Exercising the breaker can free binding mechanisms.

d) *Breaker testing*. Many breakers are fitted with thermal-magnetic trip units or magnetic (instantaneous) trip units only. Some modern MCCBs and insulated-case circuit breakers (ICCBs) on the market today contain electronic trip devices (ETDs).

    1) *Thermal unit test*. The MCCB is tested by passing 300% of the MCCB's ampere rating through each pole of the MCCB. The test results should be compared to the manufacturer's time-current characteristics (curve sheet). For multipole MCCBs, these curve sheets are based upon current in all poles of the breaker

and are utilized for coordination purposes; thus, the curve sheet should be examined for maximum single-pole trip time. Not all curve sheets specify a maximum single-pole trip time, but when one is available it should be noted.

2) *Magnetic unit (instantaneous) test.* The MCCB is tested by passing the magnetic rated trip amperes through each pole of the circuit breaker. The circuit breaker should trip within the following parameters according to the applicable frame size:

   250 A or less: +40% to –25% of the instantaneous current setting.

   400 A or greater: +25% to –25% of the instantaneous current setting.

   The testing of instantaneous trip units usually requires a high current test set for all but the smallest frame sizes of MCCBs. Care must be taken when checking the instantaneous unit with high current to insure that the thermal-trip unit is not the cause of the MCCB tripping. The high current should be placed on the MCCB for very short periods of time with adequate cool down time allowed between applications.

3) *Shunt trip.* Shunt-trip devices are utilized to trip an MCCB via some external device operation (e.g., ground-fault relay). If an MCCB is equipped with a shunt-trip coil (solenoid), the unit can be verified by applying the rated voltage across the coil, with the MCCB closed. The shunt-trip device trips a mechanical latch (or trip mechanism) that trips the MCCB.

4) *Insulation resistance.* All poles should be tested in accordance with standard insulation resistance testing guidelines at 1000 V dc. Resistance values of less than 1 MΩ should be considered inadequate and the cause should be investigated. Insulation tests should be performed between the line and load terminals with the MCCB open, between adjacent poles, and from each pole to the grounded parts of the MCCB.

5) *Contact resistance.* Contact resistance should be measured using an ohmmeter capable of measurements into the micro-ohm range. Contact resistance measurements on some of the smaller sized MCCBs are not practical where the test lead clip is larger than the MCCB's terminal.

6) *Rated load test.* A rated load hold-in test can be run if there is some doubt of the MCCB's ability to carry rated load. With the MCCB in free air, all three poles are connected in series with jumpers of a short length and adequate capacity. Applying rated load current for a minimum of 30 min should not cause the breaker to trip.

7.4 Maintenance of LVPCBs

Maintenance performed on LVPCBs should follow the manufacturer's inspection and maintenance instructions whenever possible. These instructions are detailed to the particular breaker and usually provide drawings, photographs, or sketches to point out items that must be maintained. Inspection of LVPCBs should follow a maintenance schedule established for the particular site. The first inspection should occur after an initial operating period as prescribed by the manufacturer. A thorough inspection should also be performed after the LVPCB has interrupted a short-circuit current.

Whenever a maintenance inspection procedure is being performed on an LVPCB, the following specific points of inspection should be performed:

a) Operate the LVPCB several times to make certain that the circuit breaker operates freely. Operate both manually and electrically, if so equipped, to be sure all electrical components function properly.

b) Clean all dust and dirt from the LVPCB.

c) Remove and inspect the arc chutes for any cracking, breakage, or extensive burning that would indicate a need for replacement.

d) Check the condition of contacts, both moving and stationary, and ensure proper contact penetration.

e) Check the latch mechanisms and their engagements, both open and closed.

f) Check the lubrication and lubricate as necessary per the manufacturer's instructions.

g) Check the operation of manual tripping devices.

h) Test the LVPCB to ascertain that it performs on the representative time-current coordination curves.

These tests and testing methods are described in the manufacturer's inspection and maintenance instructions. Just as in testing the MCCBs, the availability of a high current test set is necessary to completely test the LVPCB. The following tests should be performed on each pole of the LVPCB:

— Long-time delay overcurrent that protects against a circuit overload condition.
— Short-time delay overcurrent that protects for lower order fault-current conditions.
— Instantaneous-trip operation that provides circuit protection for short circuit or higher order fault-current conditions.
— Ground-fault protection that provides circuit protection for a ground-fault condition. (Ground-fault protection is usually found only on LVPCBs equipped with an ETD. The ground-fault unit will function during single-pole high-current testing of the breaker, and must be disabled per the manufacturer's instructions in order to verify the phase-trip unit.)

Many LVPCBs of modern manufacture are equipped with ETDs. Most of the manufacturers provide test units for use on their ETDs. The same performance verification testing should be performed on ETDs as those outlined above. Use of the high-current test set will test the current sensing devices on the LVPCB at the same time as the ETD if verification is desired. Generally the high-current test is not necessary for routine field maintenance checks of the ETD unless there is some question as to the sensor's operation. The test units for the manufacturer's ETDs will verify the trip characteristics of the selected settings.

7.5 Documenting maintenance results

Keeping records of maintenance work and troubleshooting can seem like an unnecessary burden. However, good maintenance records, when properly documented and analyzed, will save you time, money, and aggravation.

For example, unexpected breaker trips can result in lost production for the user, causing frustration. The user may want to know why the malfunction occurred, how long repairs will take, and what will be done to prevent the breaker trip from happening again. Keeping proper records can help to address these concerns more efficiently.

In another example, a particular breaker may trip under test at a higher current or longer time than it did when initially installed. After repeating the test a second time, the breaker performs closer to its setting. This breaker may either need to be exercised more frequently, or there may be a problem with the breaker. This condition would probably need to be discussed with the manufacturer. If this condition weren't noticed, the wire or equipment that the breaker was intended to protect could be seriously damaged if the breaker didn't trip correctly the first time as expected.

If records show that nothing is ever found wrong with a particular circuit breaker or set of breakers, this might indicate that the length of time between inspections and maintenance can be safely increased, thus saving time and money.

In order to take full advantage of documentation, records need to be taken properly. Documentation should be made not only during scheduled maintenance, but also every time it is necessary to inspect or investigate a problem on a particular piece of equipment. The information documented should be reasonably detailed. A statement such as "checked breaker" does not provide enough detail. A better record would state "checked breaker because of reported overheating," or "overheating when I arrived—found no problem." Proper documentation will help analyze and correct the equipment problem more efficiently.

7.6 Testing program

The purpose of a test procedure is to provide instructions for performing and documenting the results of tests on LVCBs. The acceptance criteria and the source of that criteria are entered on the test data record sheets immediately following the specification criteria.

7.6.1 Testing prerequisites

(Record on data record sheet. See Annexes 7A and 7B.)

a) Review the past data record sheet, and verify that previous deficiencies (if any) will not affect the test.

b) Verify that the component(s) to be tested has been released to the testing organization by the operations organization.

c) Record the current revision numbers and/or dates of implementing references used on the test data cover sheet.

d) Verify that the test procedure and the official test data recording sheet contain the latest revisions.

e) Verify that no temporary modifications will affect the performance of this procedure.

f) Notify the shift supervisor prior to commencement, stopping, or restarting this test.

g) Notify QA/QC (if so organized) of the intention to commence testing, as applicable.

7.6.2 Initial conditions

(Record on data record sheet. See Annexes 7A and 7B.)

a) Verify that work clearance tags or permits are hung on equipment to be tested, as well as on any related equipment or devices, as required to support this test.

b) Place warning signs and barriers around equipment to be tested.

c) Ensure that the test equipment in 7A.4 has a valid calibration status (if applicable). Record test equipment description, control number, and calibration due date on test data cover sheet.

7.6.3 Precautions

(Record on data record sheet. See Annexes 7A and 7B.)

a) Carefully store all equipment parts removed during the performance of this test so that they can be easily found.

b) Avoid exposing equipment to moisture, dust, dirt, or other hazardous conditions.

c) Use caution when working in or around the circuit breaker cubicles, and be aware of energized terminals and space heaters.

d) All tests shall be made only on circuit breakers that are de-energized and isolated, so that no accidental contact is made with any live parts.

e) During testing, keep hands and fingers clear of moving parts. Serious injury could result from the crushing forces that are present during power circuit breaker operation.

f) Use only manufacturer's recommended lubricants on breakers. Do not use lubricants on contact faces or on ETDs.

7.6.4 Test equipment

(Record on data record sheet. See Annexes 7A and 7B.)

a) Megohmmeter
b) Digital multimeter
c) Digital low-resistance ohmmeter

d) Circuit breaker test set
e) Psychrometer
f) Variable voltage source, ac or dc, as required
g) Appropriate gauges and tools per manufacturer's maintenance manual

7.6.5 Personnel orientation

(Record on data record sheet. See Annexes 7A and 7B.)

The lead test engineer shall ensure that personnel performing this test have been trained with respect to the operation of the electrical system, operation of the test equipment, and the electrical equipment to be tested. The lead test engineer shall also ensure that the personnel have been trained in safety precautions, data collecting techniques, and actions to be taken in the event that abnormal or unexpected conditions occur.

7.6.6 Test instructions

7.6.6.1 MCCBs (see Annex 7A)

7.6.6.1.1 Visual inspection

a) Determine ambient temperature. Record on data record sheet.

b) Check that the MCCB is free of visual defects, chipping, cracks, breaks, burns, signs of overheating, and deterioration. Record this on the data record sheet.

c) Mount MCCB onto a surface that is or can be grounded (grounded plate). Record on data record sheet.

d) Perform several mechanical on-off operations.

e) Make a circuit continuity check with a digital low-resistance ohmmeter on each pole with the circuit breaker in the closed position.

7.6.6.1.2 Primary circuit insulation resistance

NOTE—Apply megohmmeter voltage for a period of 1 min or until reading is stable for test voltage selection. (Use Table 1 in 7A.6.2.) Acceptance criteria for insulation resistance should be obtained from the manufacturer's instruction book. Typical minimum criteria are 50–100 MΩ or greater. Record test voltage and acceptance criteria on data record sheet.

a) Short all auxiliary leads on the MCCB together and connect to cubicle frame. This would include leads from auxiliary contacts or from the shunt trip coil. Record on data record sheet.

b) With the MCCB open, jumper the line-side terminals together. Jumper the load-side terminals together. Ground load-side terminals to cubicle. Record on data record sheet.

c) Connect the megohmmeter to the line-side terminals and the cubicle. Record on data record sheet.

d) Apply the test voltage to the line-side terminals and record the megohmmeter reading on the data record sheet.

e) Remove the ground from the load-side terminals and put it on the line-side terminals. Record on data record sheet.

f) Connect the megohmmeter to the load-side terminals and the cubicle. Record on data record sheet.

g) Apply the test voltage to the load-side terminals and record the megohmmeter reading on the data record sheet.

h) Remove the line- and load-side jumpers and ground. Record on data record sheet.

i) Operate the MCCB to the CLOSED position.

j) Apply megohmmeter phase-to-phase. Record on data record sheet.

k) Apply megohmmeter phase-to-ground. Record on data record sheet.

l) Remove test equipment, shorts, and ground installed for 7A.6.2 from cubicle. Record on data record sheet.

7.6.6.1.3 MCCB overcurrent trip test

CAUTION—Care should be taken to limit the maximum trip time to prevent damage to the MCCB.

a) Obtain the desired time range for the type and rating of the circuit breaker being tested, from the applicable breaker setting sheet. If a breaker setting sheet is not available, use the manufacturer's curve to obtain the correct time range at 300% of the breaker's rated circuit. Record on data record sheet.

b) Make certain that the time delay tests are made in open air at an ambient temperature of 25° C (if possible) with breakers allowed to adjust to that temperature before starting overcurrent tests.

c) Conductor leads should be of the same size as specified in UL 489-1991, and properly secured.

d) The test equipment must be capable of holding the current constant over the entire test time with as little variation as possible.

e) Close the breaker and apply 300% of breaker-rated continuous current to each pole of the circuit breaker. Repeated tests on any pole should be spaced by at least 20 min; tests on adjacent poles must be spaced by at least 5 min. Record test current on data record sheet.

f) Record the breaker trip time for each pole on data record sheet.

7.6.6.1.4 MCCB shunt-trip device (if used)

a) Obtain the percentage shunt-trip coil voltage (%TCV) and the minimum trip voltage (MTV) from the MCCB manufacturer. Calculate the MTV if only %TCV is obtained. Enter on data record sheet.

b) Apply the MTV to the shunt-trip coil and observe the MCCB trips, and record on data record sheet.

7.6.6.1.5 Instantaneous overcurrent trip test

There are two common methods for this test, one being the run-up method, in which the test set current control is set at a point equal to 70% of expected tripping current when energized. After power is turned on, the current can then be increased from this 70% value to a tripping current without excessive delay which could cause the thermal unit to trip. The second method, considered preferred, is called the pulse method and requires that the test set be equipped with a pointer stop ammeter. The current is applied in short pulses of 5–10 cycles duration. [Some manufacturers recommended that the pulsed overcurrent be first applied in excess of the instantaneous trip (IT) range, prior to adjusting for calculated ranges]. The current is increased on each succeeding step until the MCCB trips. If this meets the criteria, no further tests are required. However, it is often desirable to pulse check the tripping band width. The current is then reduced to just below that band, and by repeated pulses, the pointer stop on the ammeter is adjusted until the pointer movement is barely perceptible when the current is pulsed. The current can then be raised slightly to recheck the trip point.

CAUTION—Do not exceed the maximum current values of the manufacturer's time-current curve for the MCCB's magnetic coil.

a) When the MCCB to be tested has an adjustable instantaneous trip, adjust to MAX setting. Pulse the test setting to obtain desired magnetic overcurrent (MOC). Record MOC from breaker setting sheet on data record sheet.

b) Obtain the MCCB maximum (PLUS TOLERANCE) and minimum (MINUS TOLERANCE) tolerances from the MCCB manufacturer. Typical values are shown in 7A.6.5. Enter on data record sheet. Calculate the normal tolerance values for the MOC setting of the MCCB.

c) Alternately decrease magnetic-trip adjustment and pulse test set to obtain MOC. Record the actual MOC trip obtained on the data record sheet.

d) Perform steps a) through c) for each pole to be tested.

e) Indicate on the data record sheet the final setting for each pole adjustment of the instantaneous trip device.

f) Remove MCCB from grounding plate installed in step 7A.6.1 and record on data record sheet.

g) Enter on data record sheet the completion of step 7A.6.5.

7.6.6.1.6 Installation of draw-out MCCB cubicle

a) Ensure that the MCCB is in the OPEN position. Record on data record sheet.

CAUTION—Before installing MCCB cubicles into their respective compartments, ensure that there are no protruding metal screws or sharp edges which would damage adjacent cable insulation and that the terminations on the MCCB have the correct bolts and washers properly installed.

b) Align properly and install the draw-out MCCB cubicle in appropriate compartment. Record on data record sheet.

c) Install control transformer fuses.

d) Reterminate wires that have been lifted. Record on data record sheet.

e) Verify cubicle ground continuity to ground bus. Record on data record sheet.

7.6.6.2 LVPCB (see Annex 7B)

WARNING
Due to extreme crushing forces present during the open and close cycles of the LVPCB, care must be taken to keep hands and fingers away from the moving parts of the breaker.

7.6.6.2.1 Visual inspection

a) Record the ambient temperature on the data record sheet.

b) Remove LVPCB from its compartment. Record on data record sheet.

c) Ensure the breaker to be tested is in the OPEN and SPRINGS DISCHARGED position. Record on data record sheet.

d) Remove the ETD from the breaker. Record on data record sheet (if required to facilitate testing).

e) Mark the phase identification on each of the arc chutes and remove from the breaker. Record on data record sheet.

f) Inspect wiring terminations for tightness. Record on data record sheet.

g) Measure resistance of trip coil (TC), close coil (52X), and control relay (52Y) using a digital multimeter. Record on data record sheet.

7.6.6.2.2 Breaker mechanical operation

Perform the following steps in accordance with the manufacturer's instruction manual:

a) Contact adjustment (for butt-type contacts only). Record on data record sheet.

b) Manual slow close. Record on data record sheet.

c) Primary trip latch adjustment. Record on data record sheet.

d) Tripper bar adjustment. Record on data record sheet.

e) Primary close latch adjustment. Record on data record sheet.

f) Shunt-trip device adjustment. Record on data record sheet.

g) Magnetic latch device trip adjustment. Record on data record sheet.

7.6.6.2.3 Insulation resistance

NOTE—Apply megohmmeter voltage for a period of 1 min or until reading is stable. Acceptance criteria for insulation resistance should be obtained from the manufacturer's instruction book. Typical minimum criteria are 50–100 MΩ or greater. Record test voltage and acceptance criteria on data record sheet.

a) Connect all wires of the ETD wiring harness together and to ground.

b) Place the charging motor disconnect switch in the OFF position to isolate the charging motor.

c) Close the LVPCB; perform an insulation resistance test at 1000 V dc for 1 min or until stable reading is observed. Test each pole phase-to-ground and phase-to-phase. Record the data record sheet.

d) Remove the shorting connections from the ETD wiring. Record on data record sheet.

7.6.6.2.4 Main contact continuity test

a) With breaker closed, using a digital low-resistance ohmmeter read the resistance across the main contacts (line-to-load) on each phase. Record on data record sheet.

b) Open the breaker.

7.6.6.2.5 Close coil and trip coil minimum voltage operation test

NOTE—DC control voltages for the LVPCB below are assumed to be 125 V dc. Adjust dc power supply output voltage to LVPCB specified control voltage and minimum trip voltages if using other than 125 V dc.

a) Slowly adjust the dc power supply output voltage to 100 V dc. Record on data record sheet.

b) Close the breaker using the close push button. Record on data record sheet.

c) Adjust the dc power supply output voltage to 70 V dc. Record on data record sheet.

d) With charging motor switch in the OFF position, trip the breaker. Record on data record sheet.

e) Adjust the dc power supply output voltage to 125 V dc.

7.6.6.2.6 Time delay trip test

> **WARNING**
> If test has to be terminated in an emergency situation, depress output switch, then move test set incoming circuit breaker to OFF position.

NOTE—Use the breaker setting sheet to set the test setpoints on the ETD for long-time delay, short-time delay, and instantaneous and ground-fault trips.

a) Replace the arc chutes and the ETD in the breaker. Record on data record sheet.

b) Perform the following steps in accordance with breaker test set manual and verify that the trip times are in accordance with the manufacturer's instruction manual. Record each step on data record sheet.

   1) Long-time delay trip test
   2) Short-time delay trip test
   3) Instantaneous trip test
   4) Ground-fault trip test

c) Verify that the ETD for the breaker has been tested and the final settings have been made in accordance with the breaker setting sheet. Record on data record sheet.

7.6.7 Restoration

(Record on data record sheet. See Annexes 7A and 7B.)

a) Remove test cable and modifications, as necessary. Record on data record sheet.

b) Ensure that all equipment used during the performance of this test has been disconnected and removed, and that the surrounding area has been cleaned. Record on data record sheet.

c) Ensure that all data record sheets and tables are complete. Each data blank shall be filled out or marked N/A when not applicable. Record on data record sheet.

d) Record and identify any attachment used for the conduct of this test on page 1 of the data record sheet. Record on data record sheet.

e) Ensure that work clearances are released, as necessary. Record on data record sheet.

f) On the data record sheet, enter the initials, signature, and printed name of the individuals signing or initialing steps in the performance of this test. Record on data record sheet.

g) Verify electronic-trip setting is as per low-voltage circuit breaker setting sheet and is sealed with lead seal (if available). Record on data record sheet.

h) Verify that calculations, if any, are entered in the test data record sheet with the acceptance criteria. Record on data record sheet.

i) Ensure LVCB is completely reassembled. Record on data record sheet.

j) Make sure LVCB is left in the OPEN position. Record on data record sheet.

7.6.7.1 Breaker primary circuit to bus connections

a) With the LVPCB control power circuit de-energized, rack the LVPCB into the CONNECT position. Record on data record sheet.

b) Rack the LVPCB out to the DISCONNECT position. Record on data record sheet.

c) Verify the primary circuit connections make up evenly observing the tracks in the lubrication. Record on data record sheet.

NOTE—If an uneven primary contact makeup is displayed, adjust the contact fingers to achieve a satisfactory contact mating.

7.7 Failures detected

Should the LVCB fail any of the tests in the sequence, the tests should be stopped at that point, and the failure noted on the data record sheet. Similarly, tests should not be performed in any other sequence, because the order of testing is to protect all concerned. The insulation test is a definite first for many reasons for both the equipment and the personnel performing the test work.

7.8 References

Numerous industry standards exist for the protective functions of the breaker, as well as for performance and insulation resistance testing. Listed below are some typical standards and manuals. Some of the references may not always be applicable.

ANSI C37.16-1988, American National Standard for Switchgear—Low-Voltage Power Circuit Breakers and AC Power Circuit Protectors—Preferred Ratings, Related Requirements, and Application Recommendations.

ANSI C37.17-1979 (Reaff 1988), American National Standard for Trip Devices for AC and General Purpose DC Low-Voltage Power Circuit Breakers.

ANSI C37.20.1-1993, IEEE Standard for Metal-Enclosed Low-Voltage Power Circuit Breaker Switchgear (ANSI).

ANSI C37.50-1989, American National Standard for Switchgear—Test Procedures for Low-Voltage AC Power Circuit Breakers Used in Enclosures.

IEEE Std 4-1995, IEEE Standard Techniques for High-Voltage Testing (ANSI).[2]

IEEE Std 43-1974 (Reaff 1991), IEEE Recommended Practice for Testing Insulation Resistance of Rotating Machinery (ANSI).

IEEE Std 62-1995, IEEE Guide for Diagnostic Field Testing of Electric Power Apparatus—Part 1: Oil Filled Power Transformers, Regulators, and Reactors (ANSI).

IEEE Std 141-1993, IEEE Recommended Practice for Electric Power Distribution for Industrial Plants (*IEEE Red Book*) (ANSI).

IEEE Std 241-1990, IEEE Recommended Practice for Electric Power Systems in Commercial Buildings (*IEEE Gray Book*) (ANSI).

IEEE Std 242-1986 (Reaff 1991), IEEE Recommended Practice for Protection and Coordination of Industrial and Commercial Power Systems (*IEEE Buff Book*) (ANSI).

IEEE Std C37.13-1990 (Reaff 1995), IEEE Standard for Low-Voltage AC Power Circuit Breakers Used in Enclosures (ANSI).

IEEE C37.100-1992, IEEE Standard Definitions for Power Switchgear (ANSI).

NEMA AB 1-1993, Molded Case Circuit Breakers and Molded Case Switches (DoD).[3]

NETA ATS-1995, Acceptance Testing Specification.[4]

NETA MTS-1993, Maintenance Testing Specification.

NFPA 70B-1994, Electrical Equipment Maintenance.[5]

NFPA 70E-1995, Electrical Safety Requirements for Employee Workplaces.

UL 489-1991, Molded-Case Circuit Breakers and Circuit-Breaker Enclosures (DoD).[6]

7.9 Bibliography

Additional information may be found in the following source:

[B1] Square D publication SD363-88-1987, Field Testing Industrial Molded-Case Circuit Breakers.

---

[2]IEEE publications are available from the Institute of Electrical and Electronics Engineers, 445 Hoes Lane, P.O. Box 1331, Piscataway, NJ 08855-1331, US7A.

[3]NEMA publications are available from the National Electrical Manufacturers Association, 1300 N. 17th St., Ste. 1847, Rosslyn, VA 22209, US7A.

[4]NETA publications are available from the National Electrical Testing Association, P.O. Box 687, 106 Stone St., Morrison, CO 80465.

[5]NFPA publications are available from Publications Sales, National Fire Protection Association, 1 Batterymarch Park, P.O. Box 9101, Quincy, MA 02269-9101, US7A.

[6]UL standards are available from Global Engineering, 1990 M Street NW, Suite 400, Washington, DC, 20036, US7A.

Annex 7A

(informative)

MCCB data record

(SHEET 1 OF 6)

PERFORMED BY:
INITIAL_____DATE_____

1. UNIT ____________________ 2. EQUIPMENT TAG NO.____________ 3. ITEM NO.________

EQUIPMENT UNDER TEST

4. EQUIPMENT NAME/DESCRIPTION________________________________

5. MODEL/TYPE____________ 6. MANUFACTURER______________
7. SERIAL NO. ____________ 8. SPEC. NUMBER__________________
9. LOOP________________ 10. CLASS ______________ 11. SCHEME______________
12. FEEDER BREAKER________________

13. IMPLEMENTING REFERENCES

P&ID: ______________________________ REV: ____________________
ELEM: ______________________________ REV: ____________________
SINGLE LINE: ________________________ REV: ____________________
OTHER: __________________ REV: ____________________
OTHER: __________________ REV: ____________________
OTHER: __________________ REV: ____________________

14. TEST EQUIPMENT

DESCRIPTION	CONTROL NUMBER	CALIBRATION DUE DATE

15. ATTACHMENTS

____
____
____
____
____

(SHEET 2 OF 6)

PERFORMED BY:
INITIAL_____DATE_____

7A.1 PREREQUISITES

a) REVIEW PAST DATA RECORDS____________________________
b) COMPONENT RELEASED BY OPERATIONS FOR TESTING_____________
c) VERIFY REFERENCES ARE LATEST REVISION_____________
d) OFFICIAL TEST COPY VERIFIED_____________
e) TEMPORARY MODIFICATION EVALUATED (IF APPLICABLE)____________
f) SHIFT SUPERVISOR NOTIFIED_____________
g) QA/QC NOTIFIED_____________

7A.2 INITIAL CONDITIONS

a) WORK CLEARANCE TAGS HUNG_____________
b) WARNING SIGNS AND BARRIERS ERECTED_____________
c) TEST EQUIPMENT CALIBRATION STATUS VALID_____________

7A.3 PRECAUTIONS UNDERSTOOD_____________
7A.4 TEST EQUIPMENT ACQUIRED_____________
7A.5 PERSONNEL ORIENTATION CONDUCTED_____________
7A.6 TEST INSTRUCTIONS
7A.6.1 VISUAL INSPECTION

a) AMBIENT TEMPERATURE____________________ °F
b) MCCB FREE OF VISUAL DEFECTS_______________
c) MCCB MOUNTED ON GROUNDED PLATE____________
d) MECHANICAL ON-OFF OPERATIONS PERFORMED___________
e) CONTINUITY CHECK EACH POLE______________________

7A.6.2 PRIMARY CIRCUIT INSULATION RESISTANCE

TABLE 1

FIELD VOLTAGE	MEGOHMMETER-APPLIED VOLTS
125 Vdc	500 Vdc
240 Vac	500 Vdc
480 Vac	1000 Vdc

APPLIED VOLTAGE________________________Vdc
ACCEPTANCE CRITERIA
SOURCE_____________ MΩ___________

PROCEDURE

a) AUX LEADS CONNECTED TO CUBICLE FRAME____________
b) LINE-SIDE TERMINALS JUMPERED___________
LOAD-SIDE TERMINALS JUMPERED___________
LOAD-SIDE TERMINALS GROUNDED____________
c) MEGOHMMETER CONNECTED LINE-SIDE TO CUBICLE FRAME____________
d) LINE-SIDE TERMINALS TO CUBICLE ____________ MΩ

ACCEPTANCE CRITERIA MET____________________________________/___________
SIGNATURE DATE

(SHEET 3 OF 6)

PERFORMED BY:
INITIAL_____DATE_____

e) JUMPER FROM LOAD-SIDE TO GROUND REMOVED
LINE-SIDE TERMINALS GROUNDED___________________________

f) MEGOHMMETER CONNECTED LOAD-SIDE TO CUBICLE FRAME___________
g) LOAD-SIDE TERMINALS TO CUBICLE____________MΩ

ACCEPTANCE CRITERIA MET__________________________________/___________
SIGNATURE DATE

h) LINE-SIDE, LOAD-SIDE, AND GROUND JUMPERS REMOVED_________________
i) OPERATE THE MCCB TO CLOSED POSITION________________
j) INSULATION RESISTANCE PHASE-TO-PHASE

PHASE A – B_____________MΩ
PHASE B – C_____________MΩ
PHASE C – A_____________MΩ

ACCEPTANCE CRITERIA MET__________________________________/___________
SIGNATURE DATE

k) INSULATION RESISTANCE PHASE-TO-GROUND

PHASE A – GROUND_____________MΩ
PHASE B – GROUND_____________MΩ
PHASE C – GROUND_____________MΩ

ACCEPTANCE CRITERIA MET__________________________________/___________
SIGNATURE DATE

l) TEST EQUIPMENT, SHORTS, AND GROUNDS REMOVED______________

(SHEET 4 OF 6)

PERFORMED BY:
INITIAL_____DATE_____

7A.6.3 MCCB OVERCURRENT TRIP TEST

a) BREAKER TRIP TIME_________s _________s (ACCEPTANCE CRITERIA)
MAX MIN

SOURCE: BREAKER SETTING SHEET NO._____ CURVE NO._____

b) BREAKER TEST CURRENT

PHASE A__________ PHASE B________ PHASE C_______

c) BREAKER TRIP TIME (in seconds)

PHASE A__________ PHASE B________ PHASE C_______

ACCEPTANCE CRITERIA MET______________________________________/___________
SIGNATURE DATE

7A.6.4 MCCB SHUNT-TRIP DEVICE

a) SHUNT-TRIP COIL VOLTAGE (TCV)_____________ V
PERCENTAGES SHUNT-TRIP COIL VOLTAGE (%TCV)________%
MINIMUM TRIP VOLTAGE (MTV)_____________ V
MTV=%TCV/100 ×TCV
MTV=__________%/100 × ________ V =__________ V

b) MCCB TRIPPED____________________

(SHEET 5 OF 6)

PERFORMED BY:
INITIAL_____DATE_____

7A.6.5 INSTANTANEOUS OVERCURRENT TRIP TEST

a) MOC: FROM BREAKER SETTING SHEET_______________
b) MCCB MAX TRIP TOLERANCE (PLUS TOL)_______________
FRAME SIZE 250 A OR LESS +40%
FRAME SIZE 400 A AND GREATER +25%

MCCB MIN TRIP TOLERANCE (MINUS TOL)_____________
ALL FRAME SIZES –25%
CALCULATE THE MAX AND MIN FOR THE FINAL MAGNETIC ADJUSTMENT SETTING.
RECORD THE FOLLOWING DATA:

MAX = MOC + (MOC × PLUS TOL)

MAX =_______ + (_______ × _______) = _______ A

RECORD MAX FINAL MAGNETIC ADJUSTMENT SETTING IN THE "FINAL INSTANTANEOUS OVERCURRENT TRIP DATA BLOCK—ACCEPTANCE CRITERIA."

MIN = MOC - (MOC × MINUS TOL)

MIN =_______ + (_______ × _______) = _______ A

RECORD MIN FINAL MAG ADJ SETTING IN THE "FINAL INSTANTANEOUS OVERCURRENT TRIP DATA BLOCK— ACCEPTANCE CRITERIA."

c) FINAL INSTANTANEOUS OVERCURRENT TRIP DATA:

	ACCEPTANCE CRITERIA	ACTUAL MOC TRIP VALUES		
		PHASE A	PHASE B	PHASE C
MIN A	_________	_________	_________	_________
MAX A	_________	_________	_________	_________

ACCEPTANCE CRITERIA MET_________________________/___________
SIGNATURE DATE

d) FINAL AS LEFT SETTING PHASE A _______ PHASE B _______ PHASE C _______ A

e) GROUNDING PLATE REMOVED_______________

f) SECTION COMPLETE _________________________/___________
SIGNATURE DATE

(SHEET 6 OF 6)

PERFORMED BY:
INITIAL_____DATE_____

7A.6.6 INSTALLATION OF CUBICLE

a) BREAKER OPEN________________

b) CUBICLE IS INSTALLED________________

c) CONTROL FUSES INSTALLED________________

d) LIFTED WIRES RETERMINATED________________

e) GROUND CONTINUITY VERIFIED________________

7A.7 RESTORATION

a) TEST MODIFICATION REMOVED________________

b) TEST EQUIPMENT DISCONNECTED________________

c) DATA RECORD SHEETS COMPLETE________________

d) RECORD ATTACHMENTS USED________________

e) WORK CLEARANCES RELEASED________________

f) INDIVIDUALS SIGNING OR INITIALING STEPS________________

g) SETTING PER RELAY SETTING SHEET SEALED________________

h) CALCULATIONS ENTERED________________

i) BREAKER COMPLETELY REASSEMBLED________________

j) BREAKER IN THE OPEN POSITION________________

	SIGNATURE	PRINTED NAME	INITIAL	DATE
PERFORMED BY:				
REVIEWED BY:				
APPROVED BY:				
QA/QC REVIEW:				

Annex 7B

(informative)

LVPCB data record

(SHEET 1 OF 5)

PERFORMED BY:
INITIAL_____DATE_____

1. UNIT ____________________ 2. EQUIPMENT TAG NO.______________ 3. ITEM NO.________

EQUIPMENT UNDER TEST

4. EQUIPMENT NAME/DESCRIPTION_____________________________________

5. MODEL/TYPE______________ 6. MANUFACTURER________________
7. SERIAL NO. _______________ 8. SPEC. NUMBER____________________
9. LOOP____________________ 10. CLASS ________________ 11. SCHEME________________
12. FEEDER BREAKER__________________

13. IMPLEMENTING REFERENCES

P&ID: ______________________________ REV: ________________________
ELEM: ______________________________ REV: ________________________
SINGLE LINE: _________________________ REV: ________________________
OTHER: ____________________ REV: ______________________
OTHER: ____________________ REV: ______________________
OTHER: ____________________ REV: ______________________

14. TEST EQUIPMENT

DESCRIPTION	CONTROL NUMBER	CALIBRATION DUE DATE

15. ATTACHMENTS

____
____
____
____
____

(SHEET 2 OF 5)

PERFORMED BY:
INITIAL_____DATE_____

7B.1 PREREQUISITES

a) REVIEW PAST DATA RECORDS____________________________
b) COMPONENT RELEASED BY OPERATIONS FOR TESTING_____________
c) VERIFY REFERENCES ARE LATEST REVISION_____________
d) OFFICIAL TEST COPY VERIFIED_____________
e) TEMPORARY MODIFICATION EVALUATED (IF APPLICABLE)____________
f) SHIFT SUPERVISOR NOTIFIED_____________
g) QA/QC NOTIFIED_____________

7B.2 INITIAL CONDITIONS

a) WORK CLEARANCE TAGS HUNG_____________
b) WARNING SIGNS AND BARRIERS ERECTED________________
c) TEST EQUIPMENT CALIBRATION STATUS VALID_____________

7B.3 PRECAUTIONS UNDERSTOOD______________
7B.4 TEST EQUIPMENT ACQUIRED______________
7B.5 PERSONNEL ORIENTATION CONDUCTED______________
7B.6 TEST INSTRUCTIONS
7B.6.1 VISUAL INSPECTION

a) AMBIENT TEMPERATURE_____________°F
b) REMOVE BREAKER FROM COMPARTMENT_______________
c) ENSURE BREAKER OPEN, SPRINGS DISCHARGED_______________
d) REMOVE ETD FROM BREAKER (IF REQUIRED)_______________
e) MARK AND REMOVE ARC CHUTES_______________
f) WIRING TERMINATIONS TIGHT_______________
g) TRIP COIL (TC), CLOSE COIL (52X), AND CONTROL RELAY (52Y) RESISTANCE

RESISTANCE	NOMINAL VALVE
TC________________Ω	98 ± 10% (88–108)
52X_______________Ω	180 ± 10% (162–198)
52Y_______________Ω	2310 ± 10% (2079–2541)

SOURCE:__________________________________

ACCEPTANCE CRITERIA MET____________________________________/___________
SIGNATURE DATE

(SHEET 3 OF 5)

PERFORMED BY:
INITIAL_____ DATE_____

7B.6.2 BREAKER MECHANICAL OPERATION

a) CONTACT ADJUSTMENT____________________
b) MANUAL SLOW CLOSE____________________
c) PRIMARY TRIP LATCH ADJUSTMENT____________________
d) TRIPPER BAR ADJUSTMENT____________________
e) PRIMARY CLOSE LATCH ADJUSTMENT____________________
f) SHUNT-TRIP DEVICE ADJUSTMENT____________________
g) MAGNETIC LATCH DEVICE TRIP ADJUSTMENT____________________

7B.6.3 INSULATION RESISTANCE

a) SHORT ALL ETD WIRES TOGETHER AND GROUND (IF APPLICABLE)______________
b) ISOLATE CHARGING MOTOR____________________
c) INSULATION RESISTANCE
APPLIED VOLTAGE____________________ V dc
ACCEPTANCE CRITERIA
SOURCE________________ MΩ________________

APPLIED VOLTAGE

FROM	TO	MΩ
PHASE A	PHASE B	________________
PHASE B	PHASE C	________________
PHASE C	PHASE A	________________
PHASE A	GRD	________________
PHASE B	GRD	________________
PHASE C	GRD	________________

ACCEPTANCE CRITERIA MET________________________/___________
SIGNATURE DATE

d) SHORTING CONNECTION REMOVED FROM ETD WIRING______________

7B.6.4 MAIN CONTACT CONTINUITY TEST

a) MAIN CONTACT CONTINUITY
PHASE A____________MΩ
PHASE B____________MΩ
PHASE C____________MΩ
b) BREAKER OPEN________________

7B.6.5 CLOSE AND TRIP COILS MINIMUM VOLTAGE OPERATION TEST

(SEE NOTE IN 7.6.6.2.5)

a) 125 V dc SUPPLY ADJUSTED TO 100 V dc___________
b) BREAKER CLOSES USING CLOSE PUSHBUTTON___________
c) 125 V dc SUPPLY ADJUSTED TO 70 V dc___________
d) BREAKER OPENS___________
e) 125 V dc SUPPLY ADJUSTED TO 125 V dc___________

(SHEET 4 OF 5)

PERFORMED BY:
INITIAL_____ DATE_____

7B.6.6 TIME DELAY TRIP TEST

a) ARC CHUTES AND ETD REPLACED_____________

b) BREAKER TRIP TEST

1) LONG-TIME DELAY TRIP TEST

PHASE A___________ A ___________ s
PHASE B___________ A ___________ s
PHASE C___________ A ___________ s
ACCEPTANCE TOLERANCE (__________ s TO__________ s)
ACCEPTANCE CRITERIA: TIME CURRENT CURVE___________
SOURCE:_____________
ACCEPTANCE CRITERIA MET___________________________/___________
SIGNATURE DATE

2) SHORT-TIME DELAY TRIP TEST

PHASE A___________ A ___________ s
PHASE B___________ A ___________ s
PHASE C___________ A ___________ s
ACCEPTANCE TOLERANCE (__________ s TO__________ s)
ACCEPTANCE CRITERIA: TIME CURRENT CURVE___________
SOURCE:_____________
ACCEPTANCE CRITERIA MET___________________________/___________
SIGNATURE DATE

3) INSTANTANEOUS TRIP TEST

PHASE A___________ A ___________ s
PHASE B___________ A ___________ s
PHASE C___________ A ___________ s
ACCEPTANCE TOLERANCE (__________ s TO__________ s)
ACCEPTANCE CRITERIA: TIME CURRENT CURVE___________
SOURCE:_____________
ACCEPTANCE CRITERIA MET___________________________/___________
SIGNATURE DATE

4) GROUND FAULT TRIP TEST

PHASE A___________ A ___________ s
PHASE B___________ A ___________ s
PHASE C___________ A ___________ s
ACCEPTANCE TOLERANCE (__________ s TO__________ s)
ACCEPTANCE CRITERIA: TIME CURRENT CURVE___________
SOURCE:_____________
ACCEPTANCE CRITERIA MET___________________________/___________
SIGNATURE DATE

c) VERIFY ETD DEVICE TESTED AND FINAL SETTINGS MADE_____________

(SHEET 5 OF 5)

PERFORMED BY:
INITIAL_____ DATE_____

7B.7 RESTORATION

a) TEST MODIFICATION REMOVED____________________

b) TEST EQUIPMENT DISCONNECTED____________________

c) DATA RECORD SHEETS COMPLETE____________________

d) RECORD ATTACHMENTS USED____________________

e) WORK CLEARANCES RELEASED____________________

f) INDIVIDUALS SIGNING OF INITIALING STEPS_________________

g) ETD SETTING PER RELAY SETTING SHEET SEALED_________________

h) CALCULATIONS ENTERED____________________

i) BREAKER COMPLETELY REASSEMBLED____________________

j) BREAKER IN THE OPEN POSITION____________________

7B.7.1 BREAKER PRIMARY CIRCUIT TO BUS CONNECTIONS

a) CONTROL POWER DE-ENERGIZED____________________

RACK BREAKER INTO CONNECT POSITION____________________

b) BREAKER RACKED OUT TO DISCONNECT POSITION___________
c) PRIMARY CIRCUIT CONNECTION MAKE UP EVENLY___________

	SIGNATURE	PRINTED NAME	INITIAL	DATE
PERFORMED BY:				
REVIEWED BY:				
APPROVED BY:				
QA/QC REVIEW:				

INDEX

A

B

C

D

E

F

G

H

I

L

M

N

O

P

Q

R

S

T

U

V

W

X

Z

IEEE

IEEE

IEEE

IEEE

IEEE

IEEE